Collins

AQA A-level
Biology

Ian Honeysett

Organise and **Retrieve** your **Knowledge**

Acknowledgements

The author and publisher are grateful to the copyright holders for permission to use quoted materials and images.

P35 ©Science History Images / Alamy Stock Photo

P78 © Science Photo Library

P143 © STEVE GSCHMEISSNER/Science Photo Library

P146 © Neil Setchfield / Alamy Stock

All other images © Shutterstock.com or © HarperCollins*Publishers*

All facts are correct at time of going to press.

Published by Collins

An imprint of HarperCollins*Publishers* Limited

1 London Bridge Street

London SE1 9GF

HarperCollins*Publishers*

Macken House

39/40 Mayor Street Upper

Dublin 1

D01 C9W8

Ireland

© HarperCollins*Publishers* Limited 2025

ISBN 978-0-00-876036-6

10 9 8 7 6 5 4 3 2 1

First published 2025

British Library Cataloguing in Publication Data.

A CIP record of this book is available from the British Library.

Author: Ian Honeysett

Practice papers adapted from content written by Tom Adams

Publisher: Clare Souza

Commissioning: Richard Toms

Project management and editorial: Katie Galloway

Inside concept design: Ian Wrigley

Layout: Contentra Technologies Ltd

Cover design: Sarah Duxbury

Production: Bethany Brohm

Printed in the United Kingdom by Martins the Printers

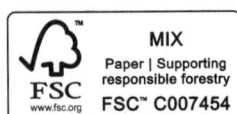

How to use this book

Use the QR code to access a support video for the topic

Each topic is presented on a two-page spread

Topics exclusive to A-level are labelled

Make your own list of key words for the topic

Answer the key questions to ensure you understand fundamental ideas

Organise your knowledge with concise explanations and examples

Make your own summary notes using your own words or sketches

Embed your knowledge with retrieval questions for each topic (use the Notes pages in the back of the book if you need more space for workings)

Specification references show you which part of the AQA course is being covered

Explore each required practical and make your own supporting notes

Try the practice papers to help prepare for the exams

Answers are provided at the back of the book (excluding for the key questions, which aim to encourage self-study)

To access the ebook, visit **collinshub.co.uk/ebooks** and follow the step-by-step instructions.

Contents

Highlighted topics/practical are A-level only

Contents

Monomers and polymers

Key words

Monomers

Monomers are small molecules that can be joined together to make larger molecules.

In carbohydrates, the monomers are monosaccharide sugars.

The most common monosaccharide sugar is glucose, $C_6H_{12}O_6$, the structure of which is shown in the diagram.

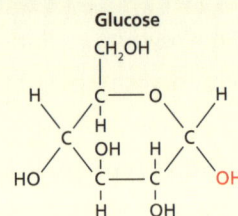

Glucose

In proteins, the monomers are amino acids.

Like carbohydrates, amino acids contain carbon, hydrogen and oxygen, but they also contain nitrogen and sometimes sulfur.

There are 20 naturally occurring amino acids. They have a similar structure but differ according to the R group. This can be a hydrogen atom or a more complicated structure, which may contain sulfur.

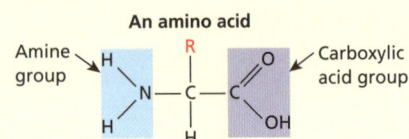

An amino acid

Amine group — Carboxylic acid group

Nucleotides are the monomers that join together to form nucleic acids such as DNA and RNA.

Nucleotides contain a nitrogenous base, a sugar unit and a phosphate group.

There are five different nitrogenous bases found in nucleic acids. Adenine, thymine, cytosine and guanine are found in DNA but RNA contains uracil instead of thymine.

In DNA, the sugar is deoxyribose sugar.
In RNA, it is ribose sugar.

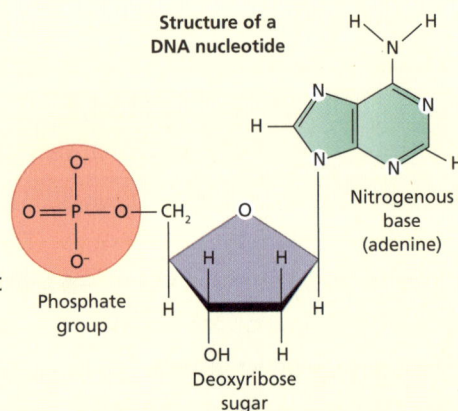

Structure of a DNA nucleotide

Nitrogenous base (adenine)

Phosphate group

Deoxyribose sugar

Forming polymers

Two monomers are joined together by a condensation reaction.

The condensation reaction forms a bond between the two monomers and involves the elimination of a water molecule.

If more and more monomers are joined together, a long chain is produced – a polymer.

The reverse of a condensation reaction is a hydrolysis reaction. This breaks the bond between monomers and involves the addition of a water molecule.

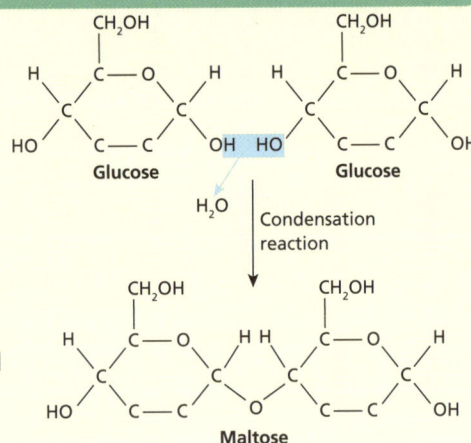

Glucose   Glucose

H_2O

Condensation reaction

Maltose

Key questions

What is an R group in an amino acid?

What components make up a nucleotide?

What is the difference between a condensation reaction and a hydrolysis reaction?

Summary

Monomers and polymers

1 The table shows details of three different monomers, A, B and C.

Monomer	Chemical formula
A	$C_{10}H_{14}N_5O_7P$
B	$C_6H_{12}O_6$
C	$C_3H_7NO_2S$

a) Identify which monomer is an amino acid, which is a monosaccharide and which is a nucleotide. Give reasons for your answers. [6]

A = ...

Reason: ....

B = ...

Reason: ....

C = ...

Reason: ....

b) The reaction between two molecules of monomer C is a condensation reaction. [1]

Give the formula of the molecule that is produced. ....

2 The diagram shows the structure of the amino acid cysteine.

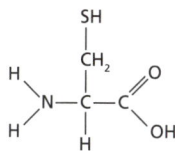

a) Draw labelled circles around three parts of the amino acid to show the positions of:

i) the R group [1]   **ii)** the carboxylic acid group [1]   **iii)** the amine group. [1]

b) In the space below, draw two cysteine molecules and show how they could combine in a condensation reaction. [2]

3 The diagram shows a nucleotide containing the base uracil.

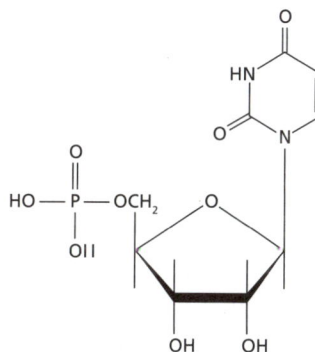

a) Draw labelled circles around three parts of the nucleotide to show the positions of the:

i) base uracil [1]   **ii)** sugar [1]   **iii)** phosphate group. [1]

b) Give **two** ways in which you can tell that this is an RNA nucleotide and not a DNA nucleotide. Use the diagram on page 6 to help you. [2]

....

....

Monosaccharides and disaccharides

Key words

Key questions

Where does the name 'carbohydrate' come from?

What is an isomer?

What is a glycosidic bond?

Monosaccharides

All carbohydrates contain carbon, hydrogen and oxygen only. The general formula for any carbohydrate is $(CH_2O)_n$

Monosaccharides are the simplest carbohydrates and are single sugar units. They can be classified according to the number of carbon atoms in the molecule.

- **Triose sugars** have three carbon atoms ($n = 3$).
- **Pentose sugars** have five carbon atoms ($n = 5$).
  Examples are ribose sugar found in RNA and deoxyribose sugar found in DNA.
- **Hexose sugars** have six carbon atoms ($n = 6$).
  This is the most common group of monosaccharides and they all have the formula $C_6H_{12}O_6$

Although hexose sugars have the same molecular formula, the atoms can be arranged in different ways. So, they can have different structures. These different molecules are called **isomers**.

Two of these isomers are α-glucose and β-glucose:

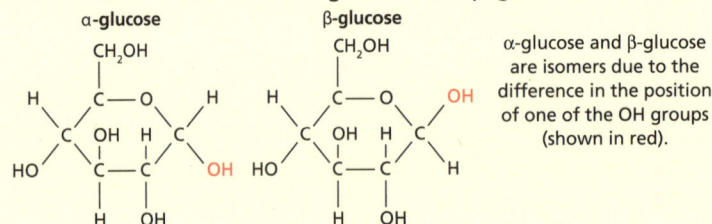

α-glucose and β-glucose are isomers due to the difference in the position of one of the OH groups (shown in red).

The structure of monosaccharides, such as α-glucose, is often drawn in a simpler form without showing all the carbon atoms in the ring or all the hydrogen atoms.

The carbon atoms are often given numbers from 1 to 6 to identify them.

Other hexose sugars include galactose and fructose.

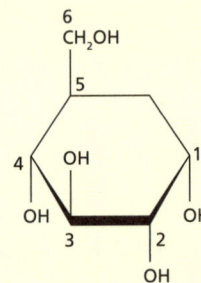

Disaccharides

Disaccharides are formed when two monosaccharides combine during a condensation reaction.

The reaction between two α-glucose molecules to form maltose is shown on page 6.

The bond formed is called a glycosidic bond.

As it is between carbon atom 1 on one glucose atom and carbon 4 on the other, it is called an α1-4 glycosidic bond.

The table shows the monosaccharides that react to form different disaccharides.

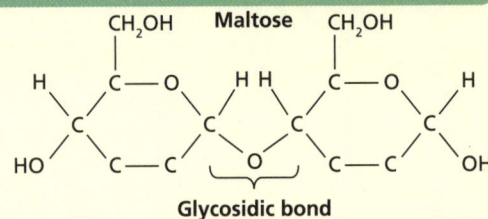

Reacting monosaccharides	Disaccharide formed
α-glucose + α-glucose	maltose
α-glucose + fructose	sucrose
α-glucose + galactose	lactose
β-glucose + β-glucose	cellobiose

Summary

Monosaccharides and disaccharides

1 The diagram shows the structure of a carbohydrate.

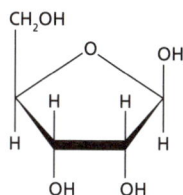

A	monosaccharide	pentose	ribose
B	disaccharide	pentose	α-glucose
C	monosaccharide	hexose	ribose
D	disaccharide	hexose	α-glucose

Which row in the table correctly describes this carbohydrate? .... [1]

2 The diagram shows two β-glucose molecules.

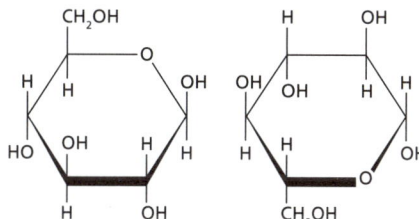

a) Show on the diagram how a glycosidic bond is formed between these two molecules. [2]

b) Give the name of the type of glycosidic bond formed. .... [1]

c) The disaccharide formed is cellobiose.

Explain why the structure of cellobiose is different from the structure of maltose shown on page 8. [2]

....

....

3 The diagram shows a molecule of sucrose.

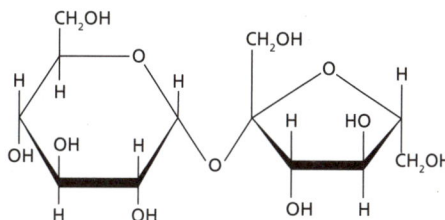

a) What is the chemical formula for sucrose? .... [1]

b) Label the glycosidic bond on the molecule. [1]

c) Name the **two** monomers that join to form sucrose. [1]

....

d) The disaccharide isomaltulose is made of the same monomers as sucrose.

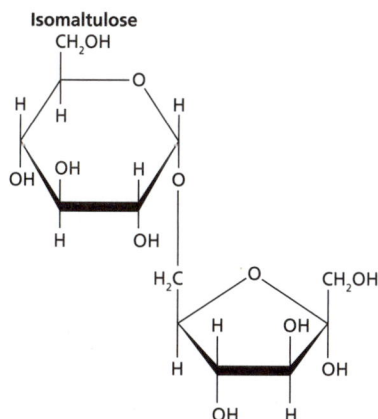

i) Label the glycosidic bond in this molecule. [1]

ii) Explain why sucrose and isomaltulose are called isomers. [1]

....

....

iii) Describe the difference in the glycosidic bond in isomaltulose compared to the bond in sucrose. [2]

....

....

....

Polysaccharides

Key words

Key questions

Which two types of molecules form starch?

Why is starch a good storage molecule?

Why is cellulose ideally suited to making cell walls?

Starch

Starch is a polymer made of long chains of α-glucose molecules joined by condensation reactions.

It contains a mixture of two different molecules:

- **Amylose** is made of a long chain of α-glucose molecules all joined by similar α1-4 glycosidic bonds. This makes the molecule wind up into a helical shape.

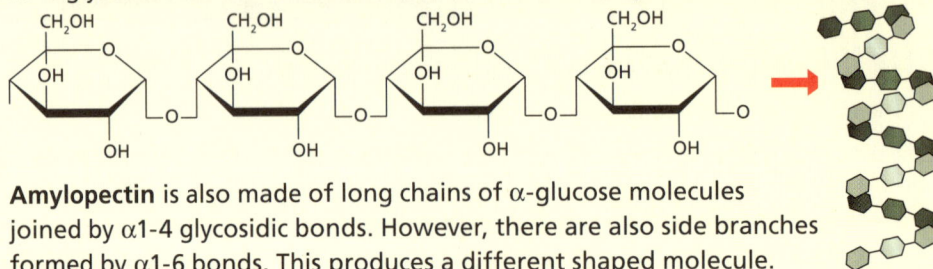

- **Amylopectin** is also made of long chains of α-glucose molecules joined by α1-4 glycosidic bonds. However, there are also side branches formed by α1-6 bonds. This produces a different shaped molecule.

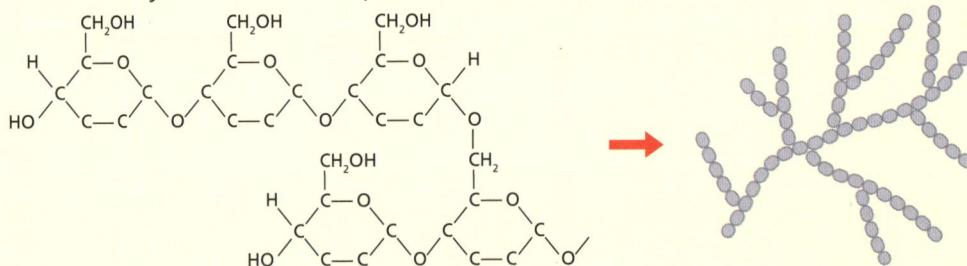

Starch acts as a store of glucose in plants. The helical shape of amylose makes it compact. The branched shape of amylopectin mean that it can be rapidly broken down from the ends of the branches by enzymes.

Glycogen

Glycogen is very similar in structure to amylopectin but with more branches. It is found in animal cells as a store of glucose.

Cellulose

Cellulose is formed of long chains of β-glucose molecules joined by β 1-4 glycosidic bonds.

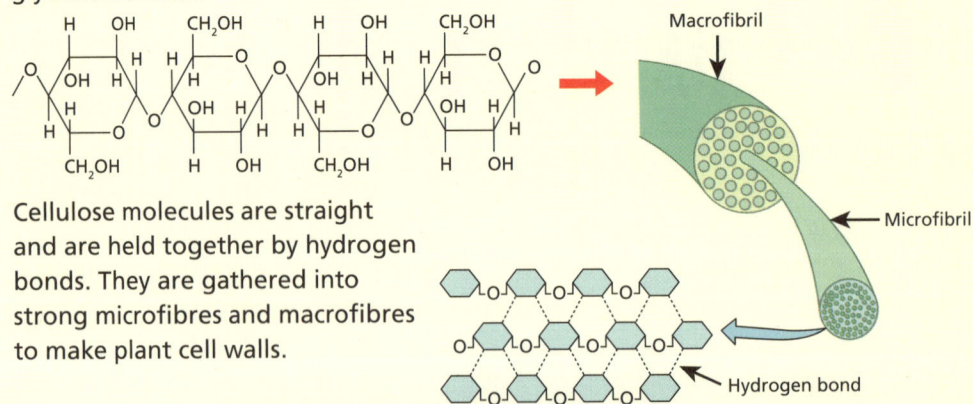

Cellulose molecules are straight and are held together by hydrogen bonds. They are gathered into strong microfibres and macrofibres to make plant cell walls.

Macrofibril

Microfibril

Hydrogen bond

Summary

Polysaccharides

(1) The table shows some features of polysaccharides.

Complete the table by putting a tick (✔) or a cross (✗) to show if these features apply to each molecule. [4]

Feature	Starch	Glycogen	Cellulose
Produced by plants			
Contains β-glucose monomers			
Contains glycosidic bonds			
Forms straight fibres			

(2) Starch is formed from a mixture of amylose and amylopectin.

Tick **one** box to show which of these statements are correct about both amylose and amylopectin. [1]

1. They contain α-glucose monomers.

2. They form branched molecules.

3. They contain only one type of glycosidic bond.

4. They act as a store of glucose.

1, 2, 3 and 4 ☐   1 and 4 only ☐   3 and 4 only ☐   2 and 3 only ☐

(3) Glycogen acts as a store of sugar.

a) Describe the shape of a glycogen molecule. [1]

....

b) Explain why the bonding in glycogen forms this shape of molecule. [2]

....

c) Explain why the structure of the glycogen molecule is suited to its storage function. [2]

....

(4) The diagram shows the structure of two polymers, chitin and cellulose.

Chitin

Cellulose

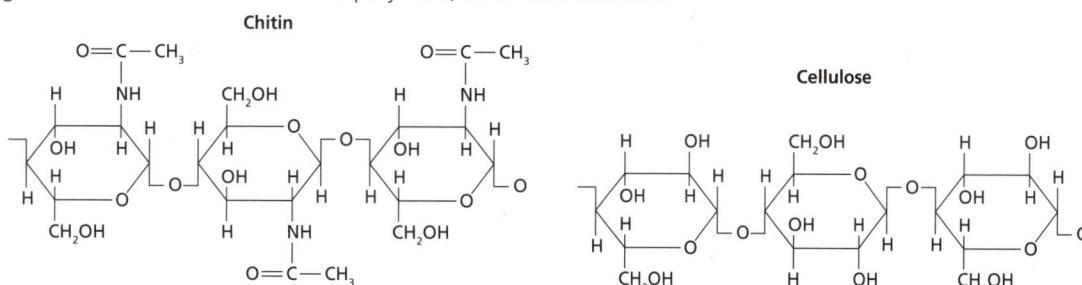

a) Name the type of glycosidic bond that links the monomers in both cellulose and chitin. .... [1]

b) Explain why chitin is not classed as a carbohydrate. [2]

....

c) Chitin forms the strong exoskeleton of insects and cellulose forms the cell walls of plants.

i) What shape of molecule is needed for both of these roles? [1]

....

ii) Explain why the type of glycosidic bond found in both of these polymers forms the shape of molecule needed for their functions. [2]

....

Lipids

General structure

Lipids are found in a wide range of forms in living organisms. They are all:

- insoluble in water
- composed of carbon, hydrogen and oxygen atoms but the ratio of oxygen to hydrogen is lower than in carbohydrates.

Triglycerides

Triglycerides are a common form of lipid and are formed from two types of molecules:

- **Fatty acids** are long chains of carbon and hydrogen atoms with a carboxylic acid group (COOH) on one end. In saturated fatty acids the long chain contains only carbon–carbon single bonds, but in unsaturated fatty acids some of the bonds are double.

Unsaturated fatty acid

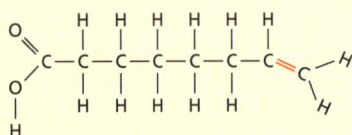

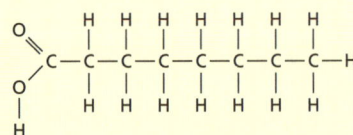

Saturated fatty acid

Saturated fatty acids have straight hydrocarbon chains but double bonds in unsaturated fatty acids cause the chain to kink.

Due to the kink, unsaturated fatty acids do not pack so closely and so have a lower melting point.

- **Glycerol** molecules are three carbon molecules with three hydroxide (OH) groups. To form a triglyceride, three fatty acids combine with one glycerol molecule in condensation reactions. The bonds that are formed are called **ester bonds**.

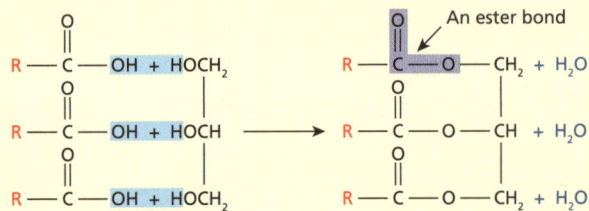

Three fatty acids   Glycerol   A triglyceride fat   Water

The R groups on the diagram represent the hydrocarbon chains. A large number of different fatty acids can be used to make a triglyceride.

Triglycerides are used as storage molecules as they are insoluble, can release large amounts of energy and are compact. They also provide thermal insulation and mechanical protection.

Phospholipids

Phospholipids are formed in a similar way to triglycerides but only contain two fatty acids. Instead of a third fatty acid, they have a phosphate group attached to the glycerol molecule. This means that one end of the molecule (the phosphate end) is attracted to water. The other end (the hydrocarbon chains) is repelled by water. This is important in the formation of cell membranes (page 38).

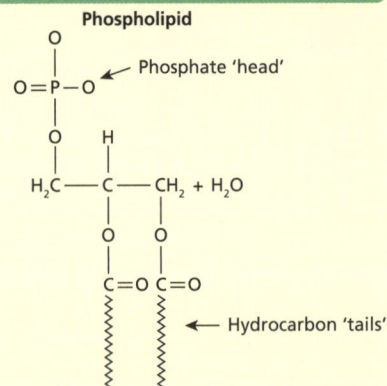

Phospholipid

← Phosphate 'head'

← Hydrocarbon 'tails'

Lipids

1 The table gives details of three different molecules.

Molecule	Chemical formula
A	$C_{12}H_{22}O_{11}$
B	$C_{55}H_{98}O_6$
C	$C_{39}H_{98}O_{10}P$

  a) Identify which molecule is a carbohydrate, which is a triglyceride and which is a phospholipid. Give reasons for your answers. [6]

  A = ...

  Reason: ....

  B = ...

  Reason: ....

  C = ...

  Reason: ....

2 Tick **one** box to show which of these statements are correct about both saturated and unsaturated fatty acids. [1]

  1. They can form an ester bond with glycerol.

  2. They have a straight hydrocarbon tail.

  3. They contain a carboxylic acid group.

  4. They contain carbon–carbon double bonds.

  1, 2, 3 and 4 ☐   1 and 3 only ☐   2 and 4 only ☐   2 and 3 only ☐

3 The diagram shows a representation of a triglyceride molecule.

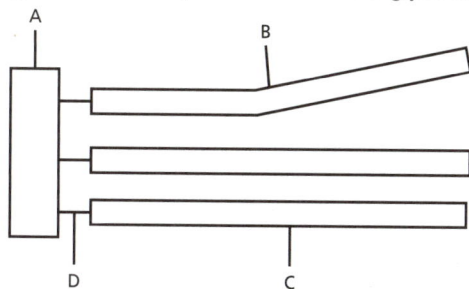

  a) Identify parts A, B, C and D. [4]

  A = ...

  B = ...

  C = ...

  D = ...

  b) Explain why triglycerides are well adapted to the function of energy storage. [3]

  ....

  ....

  ....

4 If phospholipids are in a liquid such as blood or water, they can gather together to make circular structures called micelles or liposomes.

These are shown in the diagram.

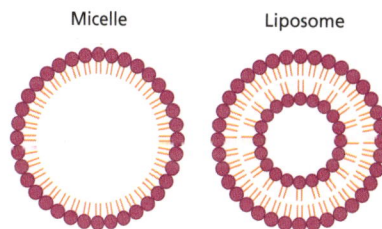

Micelle   Liposome

  a) Explain why phospholipids arrange themselves in these structures when added to blood or water. [2]

  ....

  ....

  b) One of these molecules can be used to transport lipid soluble drugs through the blood. Explain which of these structures can do this. [2]

  ....

  ....

Proteins

Key words

Key questions

How is a dipeptide formed?

Building proteins

Proteins are long chains of amino acids. Two amino acids are joined by a condensation reaction between the amine group in one amino acid and the carboxyl group in the next.

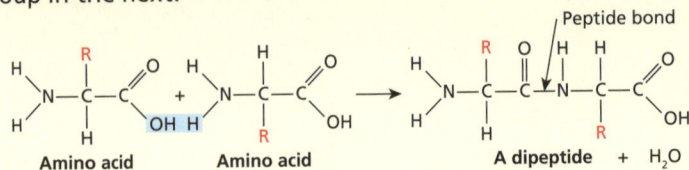

Amino acid   Amino acid   A dipeptide   + H_2O

Many more amino acids are added to form a polypeptide, which then folds up to create a protein. As there are 20 different amino acids, they can combine in many different orders to form a wide variety of proteins.

Levels of structure

Proteins are complex molecules and their structure can be described in different levels:

- **Primary structure** describes the order of amino acids in the polypeptide chain.
- **Secondary structure** describes how certain areas of the polypeptide twist or coil as it is formed. Two of the most common examples of secondary structures are the α-helix and β-pleated sheet.

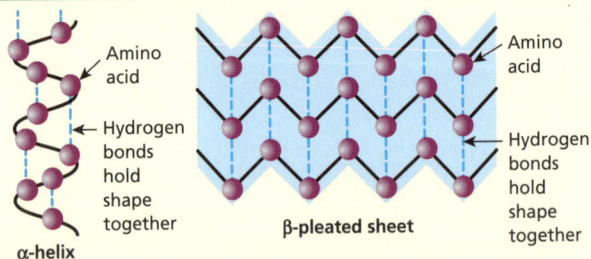

What is the primary structure of a protein?

- **Tertiary structure** describes the 3D shape formed by the whole polypeptide molecule as it forms a protein. The tertiary structure is held together by different types of attractions:
  - hydrogen bonds, which are weak attractions (see page 28)
  - ionic bonds, which are strong bonds between charged groups
  - disulphide bridges between sulfur atoms in the R groups of certain amino acids
  - hydrophobic and hydrophilic interactions, where some R groups tend to be on the outside of the molecule (near water) and others on the inside.

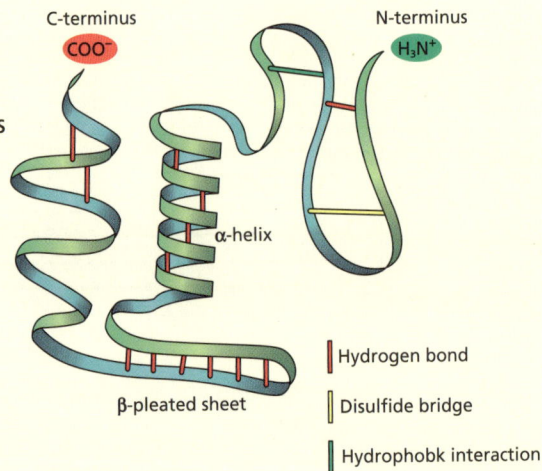

Why do only some proteins have a quaternary structure?

The order of amino acids decides which interactions occur and so the tertiary structure is decided by the primary structure.

Quaternary structure only occurs in proteins that have more than one polypeptide chain. It describes how the chains are arranged. An example is haemoglobin that has four chains, each with a non-protein haem group.

Proteins may be fibrous or they may be globular. Fibrous proteins are structural. Globular proteins can be transported in the blood, such as haemoglobin. All enzymes are globular proteins.

✔ Summary

RETRIEVE

Proteins

1 The diagram shows a molecule formed from amino acids.

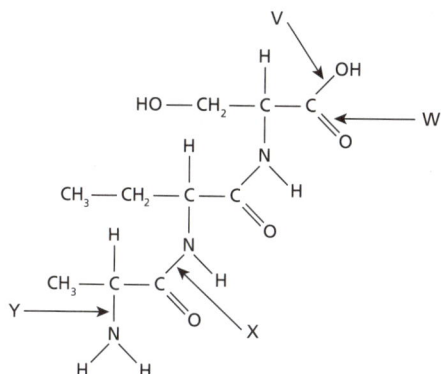

a) How many amino acids formed this molecule?

_____ [1]

b) Which letter, V, W, X or Y, labels a peptide bond?

_____ [1]

c) What is the number of atoms in the largest R group in the molecule?

_____ [1]

2 The protein haemoglobin has four polypeptide chains: two identical α chains and two identical β chains. The table shows some features used to describe protein molecules.

Number	Feature
1	The order of amino acids in the polypeptide chain
2	α-helix
3	Ionic bonds
4	β-pleated sheet
5	Peptide bonds
6	The 3D structure of each polypeptide chain
7	The way that the α chains and the β chains fit together

Give the number, or numbers, of any of the features in the table that describe each of the following.

a) The primary structure of the α chains _____ [1]

b) Any features that are examples of secondary structures _____ [1]

c) Any bonds that determine the tertiary structure _____ [1]

d) The quaternary structure of haemoglobin _____ [1]

3 The diagram shows part of a protein molecule.

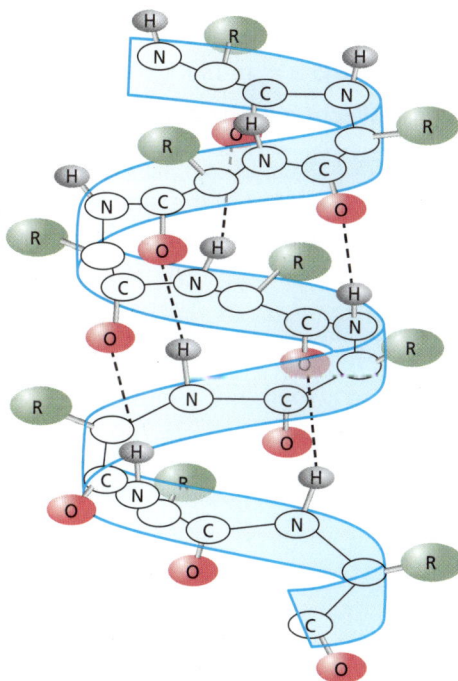

a) Name the structure shown in the diagram.

_____ [1]

b) Name the level of protein structure shown.

_____ [1]

c) Describe how this structure is held together. [2]

_____

_____

_____

d) State why this structure can be easily broken by increasing temperatures. [1]

_____

_____

Testing for biological molecules

ORGANISE

Key words

Key questions

Why are different tests needed for glucose and sucrose?

Why is hydrochloric acid used to test for sucrose?

Why is ethanol needed to test for lipids?

Carbohydrates

There are biological tests that can be used in the laboratory to test for three main categories of carbohydrates.

Reducing sugars include any sugars that can act as reducing agents. All monosaccharides such as glucose, galactose and fructose are reducing sugars. Some disaccharides, such as lactose and maltose, are reducing sugars but sucrose is a non-reducing sugar.

Testing for reducing sugars involves using Benedict's solution:

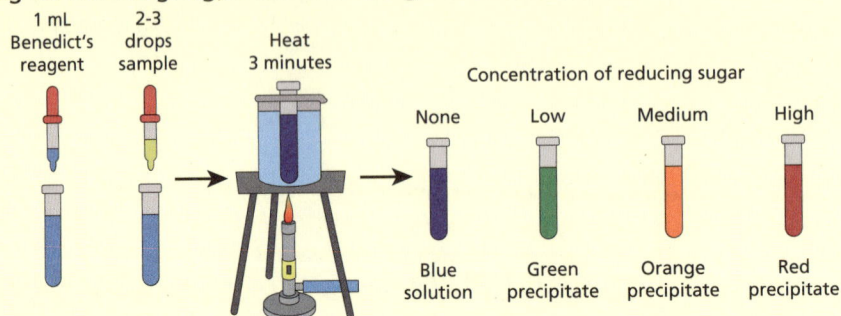

The Benedict's test is usually used as a qualitative test as a red colour indicates that the sample contains reducing sugar. (Orange or green indicate lower concentrations.) If the colour of the solution is measured, the test can be quantitative.

Non-reducing sugars include sucrose. To test for a non-reducing sugar, a Benedict's test is performed and a negative result obtained. Then the disaccharide is hydrolysed by heating in acid. The resulting monosaccharides give a positive test with Benedict's solution.

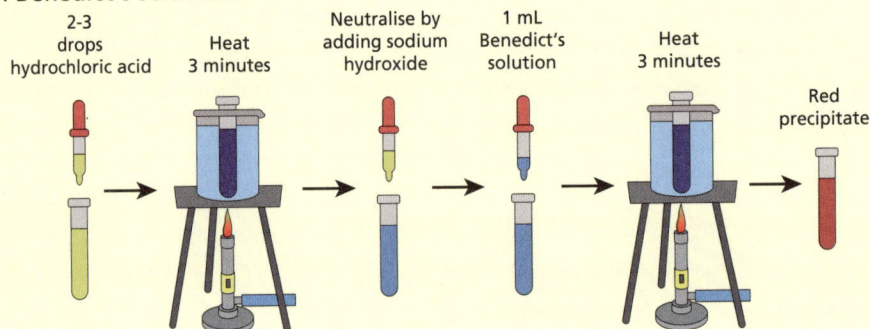

Starch is tested for using iodine solution (iodine in potassium iodide). A few drops of iodine solution are added to the sample and a blue-black colour indicates starch.

The test works due to the iodine atoms collecting in the amylose helix.

Protein

Protein is tested for using biuret reagent. When added to the sample, a purple/mauve colour indicates proteins and a blue colour is negative. The biuret reagent works by reacting with peptide bonds.

Lipids

The test for lipids is called the emulsion test. Ethanol is added to the sample to dissolve any lipid. The ethanol/lipid solution is then poured into water. Any lipid present forms a milky white emulsion.

Summary

RETRIEVE

Testing for biological molecules

1 A student tested a sample of natural yogurt for different biological molecules.

The table shows some of the student's results.

a) Complete the table. [8]

Reagents	Molecule tested for	Colour produced	Positive or negative result
iodine solution			negative
biuret reagent			positive
Benedict's solution		red precipitate	
ethanol and water			positive

b) Explain why the result for Benedict's solution shown in the table is described as qualitative. [1]

c) The result shown in the table for ethanol and water is positive.

Explain why the student is unsure of this positive result for the yogurt sample. [1]

2 The diagram shows the reaction between iodine solution and one of the components of starch.

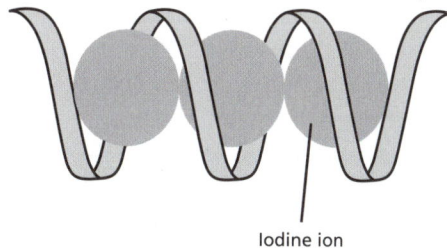

Iodine ion

a) Give the name of the component of starch shown in the diagram. [1]

b) State the colour change seen as a result of this reaction. [1]

c) Use the diagram to explain why iodine solution cannot be used to test for glycogen. [2]

d) The helical structure in the diagram is held together by hydrogen bonds.

Explain why iodine solution cannot be used to test for starch at temperatures above 60°C. [2]

3 A student has a solution containing a reducing sugar and a non-reducing sugar.

Describe how the student could show that both sugars are present in the solution. [5]

Mode of enzyme action

Key words

Key questions

How do enzymes increase the rate of reaction?

What is the lock and what is the key in Fischer's model for enzyme action?

How does the induced-fit model for enzyme action get its name?

Function of enzymes

Enzymes are globular proteins that act as biological catalysts in living organisms. They are needed because most chemical reactions would be too slow in the conditions found in living organisms.

High temperatures would increase the rate of the reactions but would destroy cells. Enzymes increase reaction rates at the lower temperatures found in living systems.

Enzyme-controlled reactions can take place inside cells (intracellular) or can act outside cells (extracellular). The reactions that enzymes catalyse may be part of metabolic processes such as respiration and photosynthesis, or may produce structural molecules needed to build cells and tissues.

Activation energy

Even if chemical reactions have a net output of energy (exothermic), they still need an initial input of energy to start the reaction. This is called the activation energy.

Enzymes speed up the rate of reactions by lowering the activation energy.

The activation energy for the reaction is reduced because enzymes can:

- bring substrates together, making them more likely to react
- exert stress on a substrate molecule, making it more likely to break up.

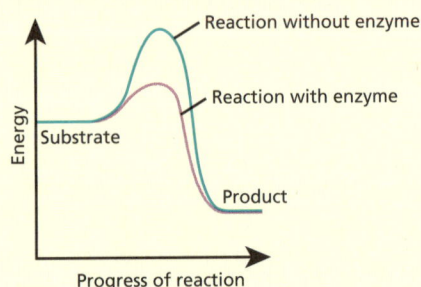

Models for enzyme action

Any model for enzyme action must explain the properties of enzymes. They:

- reduce the activation energy for a reaction
- are specific for a particular reaction
- can catalyse both the forward and reverse reaction in reversible reactions
- remain unchanged at the end of a reaction.

The first major theory for how enzymes work was put forward by Emil Fischer in 1894 – the Lock and Key model.

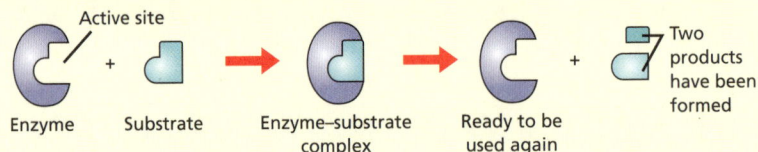

As in most fields of science, advances in technology result in the development of new models that more accurately fit data. In the case of enzyme action, it was realised that:

- different parts of the enzyme molecule move in response to the environment
- most of this movement occurred when the substrate bound with the active site.

These discoveries led to the development of the induced-fit model for enzyme action.

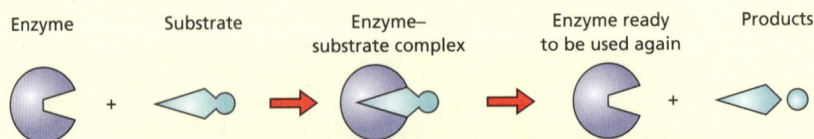

In this model the substrate interacts with the enzyme, changing the shape of the active site. This allows the active site and the substrate to fit correctly. Once the product leaves the enzyme, the active site changes back to its original shape.

Summary

Mode of enzyme action

1 Amylase is released from the salivary gland to digest amylose in the mouth.

The graph shows the energy changes during the digestion of amylose with and without amylase.

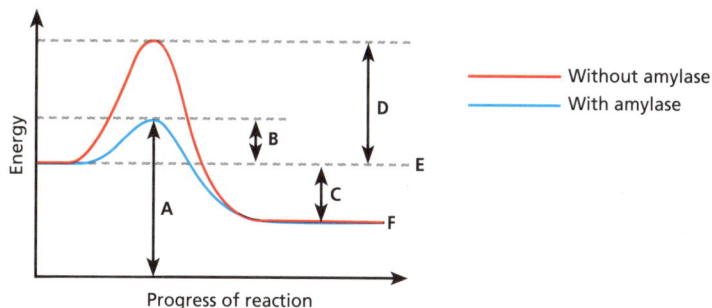

a) Why is amylase described as an extracellular enzyme? [1]

b) Identify which letter on the graph represents each of these aspects of the digestion of starch by amylase. [4]

The energy of amylose molecules. ☐   The activation energy of the reaction without amylase. ☐

The energy of maltose molecules. ☐   The activation energy of the reaction with amylase. ☐

c) Use the graph to explain why amylase increases the rate of reaction for the digestion of amylose. [2]

d) Use the graph to show that the digestion of amylose is exothermic. [2]

2 The diagram shows a model for a mechanism for enzyme action.

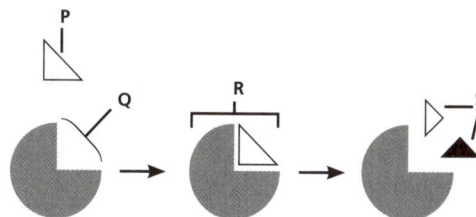

a) What names are given to P, Q, R and S in the diagram? [4]

P = _____   Q = _____

R = _____   S = _____

b) Explain the difference between this model and the induced-fit model of enzyme action. [2]

3 The enzyme ethanol dehydrogenase catalyses the conversion of ethanol to ethanal. The same enzyme can also catalyse the conversion of ethanal to ethanol.

a) Explain why ethanol dehydrogenase can catalyse both reactions. [2]

b) The diagram shows the structure of ethanol and a compound called diethylene glycol.

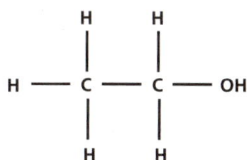

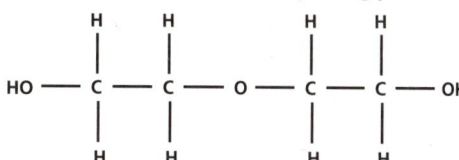

Ethanol   Diethylene glycol

Explain why ethanol and diethylene glycol can both act as a substrate for ethanol dehydrogenase. [2]

Factors affecting enzyme action

🔑 Key words

❓ Key questions

Why does denaturing an enzyme prevent it working?

What is a limiting factor?

Why are some inhibitors called 'competitive'?

Temperature and pH

Temperature can affect the rate of an enzyme reaction in two ways.

Increasing temperature gives enzyme and substrate molecules more kinetic energy, so they are more likely to collide. This increases the rate of reaction.

Above the optimum temperature, any further increase changes the bonding in the enzyme. This changes the shape of the active site so the substrate does not fit so easily. At high temperatures, the enzyme's tertiary structure is permanently disrupted and the enzyme is denatured.

Small changes in **pH** can change the charges in the active site. Larger changes can alter the tertiary structure of the enzyme and the active site. Both of these factors can reduce the rate of reaction. Different enzymes have different optimum pH values depending on where in the body they are used.

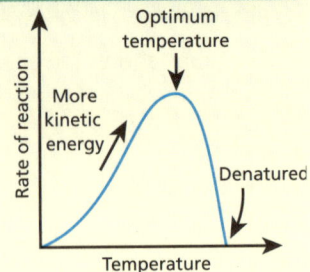

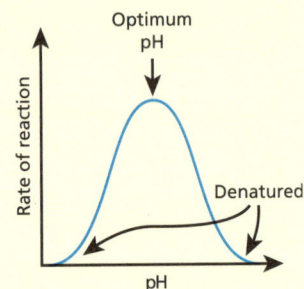

Substrate and enzyme concentration

If either of these concentrations are low, there are fewer collisions and the rate of reaction is lower.

- If the substrate concentration increases, the rate increases but levels off as the enzyme concentration becomes limiting.
- If the enzyme concentration increases, the rate increases but levels off as the substrate concentration becomes limiting.

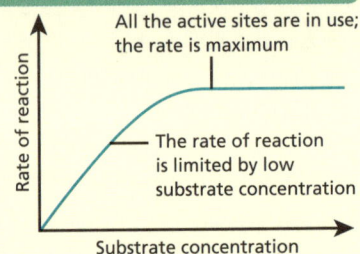

Inhibitors

Inhibitors are chemicals that slow down the rate of an enzyme catalysed reaction. There are two main types of inhibitors.

Competitive inhibitors:
- have a similar shape to the substrate and so can bind with the active site
- will not form products when in the active site
- block the site stopping the substrate entering
- are usually able to leave the active site.

Non-competitive inhibitors:
- bind to the enzyme away from the active site
- cause the active site to change shape preventing the substrate entering or reacting
- often have a permanent effect.

The effect of competitive inhibitors can be reduced by increasing the concentration of the substrate. However, the maximum rate of reaction cannot be reached if a non-competitive inhibitor is involved.

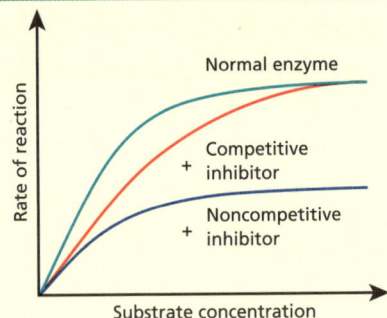

✔ Summary

Spec. ref. 3.1.4.2

Factors affecting enzyme action

1. Trypsin is an enzyme found in the small intestine. The optimum conditions for trypsin are a pH of 7.8 and a temperature of 37°C.

Put a tick next to any statement about the action of trypsin that is true. [1]

Most enzyme-substrate complexes will be formed at a pH of 7.8 and a temperature of 37°C. ☐

An increase in body temperature above 37°C will increase the activity of trypsin. ☐

A fall in pH below 7.5 will start to change the primary structure of trypsin. ☐

Below 37°C, hydrogen bonds in the trypsin molecule will start to break. ☐

2. The graph shows the results of an experiment measuring the rate of hydrolysis of maltose by maltase.

The experiment was carried out using a range of maltose concentrations and using two different maltase concentrations.

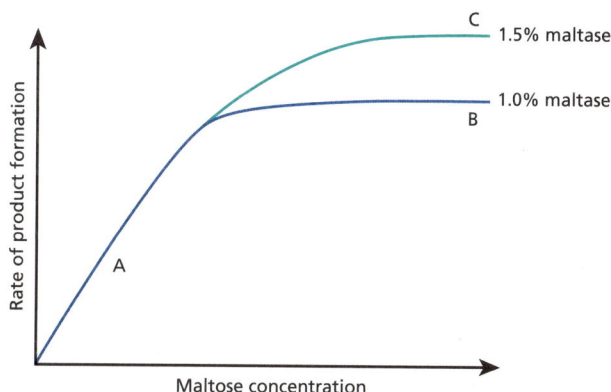

a) i) State which factor is limiting the rate at point A. _____ [1]

   ii) Explain why increasing this factor will increase the rate. [2]

b) i) State which factor is limiting the rate at point B. _____ [1]

   ii) State how you can tell which factor is limiting the rate at point B. [1]

c) Suggest a factor that could be limiting the reaction at point C. _____ [1]

3. Which of these statements are true about all enzyme inhibitors? Tick **one** box. [1]

   1. They bind to the active site of the enzyme.

   2. They reduce the chance of an enzyme-substrate complex forming.

   3. They change the shape of the active site.

   1, 2 and 3 ☐   1 and 2 only ☐   1 and 3 only ☐   2 only ☐

4. Bacteria need to produce folic acid to allow them to reproduce.

They do this using an enzyme called DHPS that acts on a molecule called PABA.

Doctors use drugs containing sulfonamide to treat bacterial infections.

The diagram shows the structure of PABA and sulfonamide.

Explain why sulphonamides can be used to treat bacterial infections. [4]

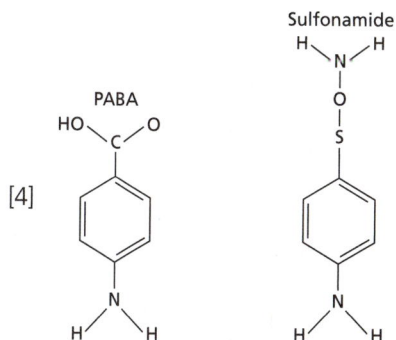

Structure of DNA and RNA

Key words

Key questions

What is a phosphodiester bond?

What is complementary base pairing?

How does RNA differ from DNA?

Structure of DNA

DNA is a polymer that is made up of long chains of nucleotides. The structure of DNA nucleotides is described on page 6. They can be drawn in simple form, as shown.

There are four different nucleotides, each containing a different base:

- adenine or guanine, which are called purines
- cytosine or thymine, which are pyrimidines.

The pentose sugar is deoxyribose sugar.

Two scientists, Watson and Crick, proposed a structure for DNA based on the four nucleotides. These are the features of their model:

- The nucleotides are joined together by condensation reactions between the phosphate groups and the pentose sugars. These reactions form phosphodiester bonds.
- The DNA molecule has two polynucleotide chains joined by hydrogen bonds between the bases. This is complementary base pairing (A with T and G with C).
- The two chains run in opposite directions – this is described as antiparallel and means that the ends of strands (called 3' and 5') are different on each strand.
- The two polynucleotide chains are wound together to form a double helix, with ten nucleotides per turn of the chain.

Simplified structure of a DNA nucleotide

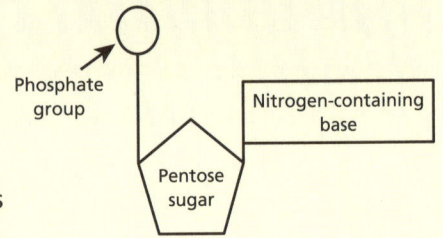

Watson and Crick model of DNA

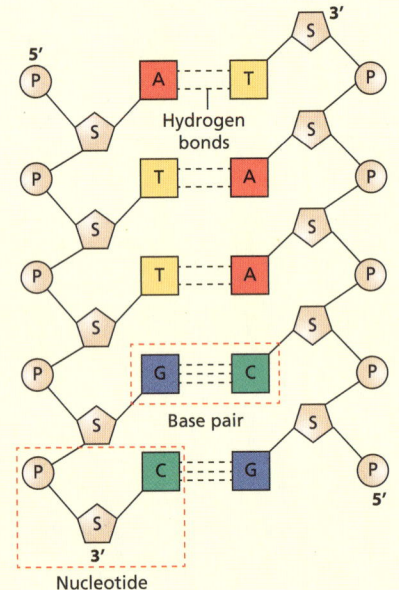

Structure of RNA

RNA is also a polymer of nucleotides but, unlike DNA, it is a single polynucleotide chain.

The pentose sugar is ribose sugar and there is no thymine base, which is replaced by uracil.

There are three main types of RNA molecule that have different shapes and functions:

- **Messenger RNA** (mRNA) is formed in the nucleus and carries the information for protein synthesis into the cytoplasm.
- **Transfer RNA** (tRNA), which carries amino acids to the ribosomes.
- **Ribosomal RNA** (rRNA), which combines with proteins to form ribosomes.

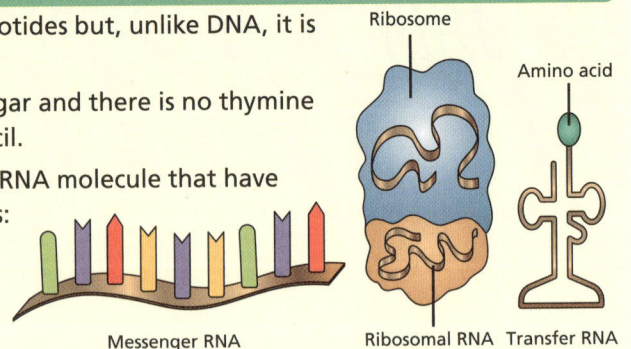

Summary

Structure of DNA and RNA

1 Complete this table to give three differences between the structure of DNA and the structure of RNA. [3]

	DNA	RNA
1		
2		
3		

2 Tick **one** box to show which of these statements are true about the bonding in DNA. [1]

1. Phosphodiester bonds bind one polynucleotide strand to the other.

2. A purine base forms hydrogen bonds with another purine base.

3. Phosphodiester bonds occur between phosphate and sugar groups.

1, 2 and 3 ☐

1 and 2 only ☐

2 only ☐

3 only ☐

3 The diagram shows a section of a DNA molecule.

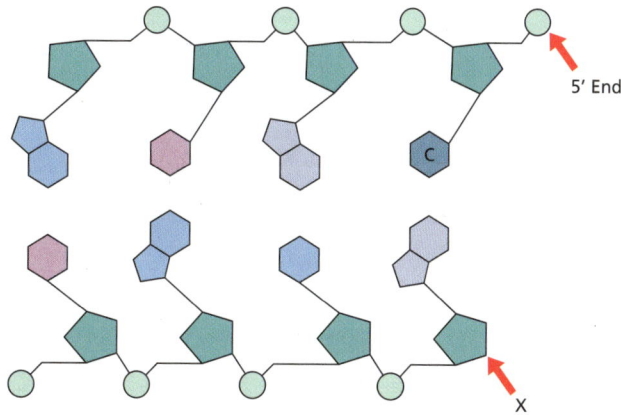

5' End

C

X

a) Draw a circle around **one** nucleotide. [1]

b) One of the bases has been labelled with a C for cytosine.

Label each of the other bases with C, T, A or G. [4]

c) Draw in all the hydrogen bonds between the bases. [2]

d) Which end of the molecule has been labelled X? [1]

4 The table shows some of the results of an experiment measuring the percentage of bases in DNA from different organisms.

Organism	%A	%G	%T	%C
Human	30.9		29.4	
Chicken	28.0	22.0	28.4	21.6
Bacterium	24.7	26.0	23.6	25.7

a) Estimate the missing percentages for humans and write them in the table. [2]

b) Comment on the pattern of results for the three species. [3]

DNA replication

🔑 Key words

❓ Key questions

What enzymes are involved in DNA replication?

Why is DNA replication called semi-conservative?

How does DNA code for proteins?

DNA replication

Any genetic material needs to carry out two functions:
- Replicate itself so that information can be passed on to the next generation.
- Code for the production of proteins.

When Watson and Crick proposed a structure for DNA, they followed it up with an explanation for how DNA replicated. They called this **semi-conservative** replication.

These are the steps in this process:
- An enzyme called helicase unwinds the double helix of the molecule and breaks the hydrogen bonds between the complementary bases. This causes the two strands to come apart or 'unzip'.
- New DNA nucleotides are attracted to the exposed bases and complementary base pairing occurs.
- Another enzyme, DNA polymerase, catalyses the formation of phosphodiester bonds to join adjacent nucleotides.
- The two new molecules then wind up.

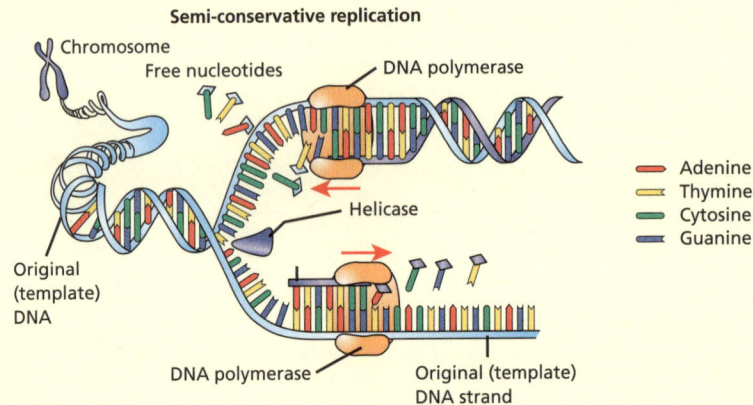

Semi-conservative replication

The replication is called semi-conservative because each new molecule contains one original strand and one new strand. Each of the original strands acts as a template for the new strands.

Validating Watson and Crick's proposals

Meselson and Stahl's experiments proved that DNA replication was semi-conservative:
- They used an isotope of nitrogen, ^{15}N, which is a heavier atom than the more common ^{14}N.
- They grew bacteria in a culture medium that contained ^{15}N for many generations so that all the DNA contained ^{15}N.
- They then transferred these bacteria to a medium containing only ^{14}N and allowed the bacteria to divide once.
- They showed that the DNA produced was intermediate in mass to DNA containing all ^{15}N and DNA containing all ^{14}N.

Therefore, the molecules must have one new ^{14}N strand and one old ^{15}N stand.

As well as suggesting semi-conservative replication, Watson and Crick proposed that DNA controlled protein structure by coding for the order of amino acids: that the order of the bases codes for the order of amino acids in a protein (see page 62). Many scientists questioned whether just four bases could code for the range of proteins produced. Further experiments by Crick proved that the base sequence was the code.

✔ Summary

DNA replication

1) DNA replication involves the enzymes DNA helicase and DNA polymerase.

Put a tick (✓) or cross (✗) in each box in the table to show the functions of these enzymes in DNA replication. [8]

	DNA helicase	DNA polymerase
Breaks hydrogen bonds		
Breaks phosphodiester bonds		
Forms phosphodiester bonds		
Unwinds DNA		

2) The DNA in one human cell contains 6.2×10^9 nucleotide pairs.

a) One nucleotide pair is 3.4×10^{-10} m long.

Calculate the length of all the DNA in one human cell. [2]

b) DNA replication takes six hours in human cells. One DNA polymerase enzyme can join 50 DNA nucleotides together per second on both DNA strands.

Calculate the number of enzyme molecules needed to replicate the DNA in one human cell in six hours. [3]

3) Meselson and Stahl investigated DNA replication, as described on page 24. They demonstrated the mass of the DNA produced by extracting it from the bacteria and putting it in a sugar solution.

The greater the mass on the DNA, the further down the test tube the DNA would settle.

The diagram shows some of the results of their experiment.

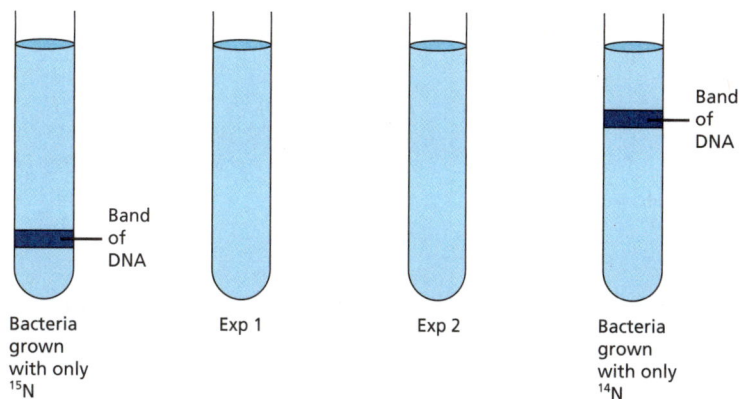

Bacteria grown with only ^{15}N — Exp 1 — Exp 2 — Bacteria grown with only ^{14}N

a) The tube labelled Exp 1 shows the results from their first experiment, where bacteria grown with only ^{15}N were allowed to divide **once** in a medium with only ^{14}N.

i) Draw on tube Exp 1 the expected results. [1]

ii) Explain the results. [3]

b) The tube labelled Exp 2 shows the results from a second experiment, where bacteria grown with only ^{15}N were allowed to divide **twice** in a medium with only ^{14}N.

i) Draw on tube Exp 2 the expected results. [1]

ii) Explain the results. [3]

ATP

Key questions

What is ATP synthase?

ATP structure

Adenosine triphosphate (ATP) is a nucleotide, similar in structure to DNA and RNA nucleotides. It is made up of the base adenine, joined to ribose sugar and three phosphate groups.

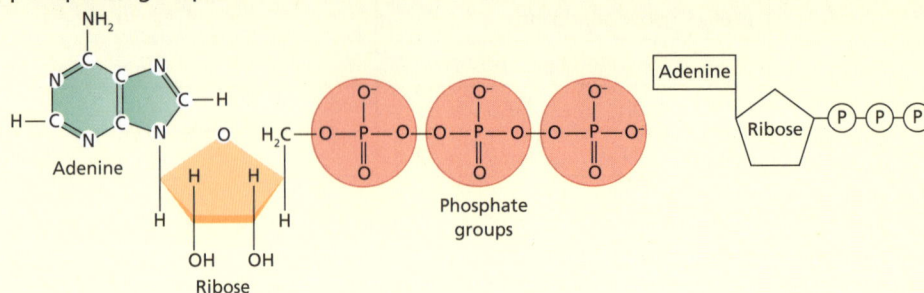

ATP is formed by a condensation reaction between adenosine diphosphate (ADP) and an inorganic phosphate group (P_i). The enzyme ATP synthase catalyses this reaction.

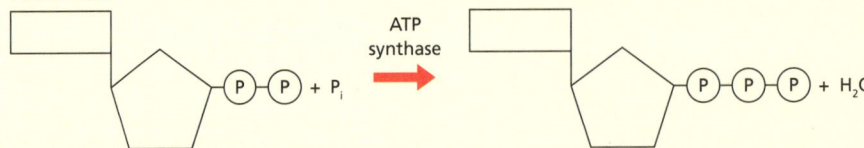

How does ATP release energy?

The reverse hydrolysis reaction breaks down ATP to ADP and P_i. This is catalysed by the enzyme ATP hydrolase.

ATP function

Metabolism is the sum of all the chemical reactions occurring in an organism. Some of these reactions, such as respiration, release energy (catabolic) and some, such as protein synthesis, need an input of energy (anabolic).

ATP acts as the link between these two sets of reactions by acting as an energy carrier.

Why is ATP called the 'universal energy carrier'?

30.6 kJ of energy is released when one mole of ATP is hydrolysed to ADP. The same amount of energy is needed to convert one mole of ADP to ATP.

The production of ATP occurs in the processes of respiration and photosynthesis.

ATP is mobile and so can move about in cells.

Cells do not store large quantities of ATP but its role is to transfer energy between different sites in the cell and between different processes. This means that cells do not need different energy transfer systems for different processes.

ATP is found in all living cells and so is often called the **universal energy carrier**.

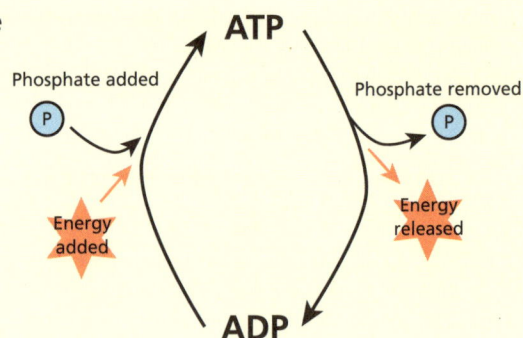

Summary

ATP

1. Which of these processes both produce ATP? Tick **one** box. [1]

Respiration and protein synthesis ☐ Photosynthesis and respiration ☐

Photosynthesis and active transport ☐ Muscle contraction and respiration ☐

2. Tick **one** box to show which of these statements about ATP are correct. [1]

   1. ATP is hydrolysed by ATP synthase.

   2. Muscle cells store large quantities of ATP.

   3. ATP can contain any one of four possible organic bases.

   2 and 3 only ☐   1 and 2 only ☐   3 only ☐   None of the statements ☐

3. The diagram shows two reactions involving ATP.

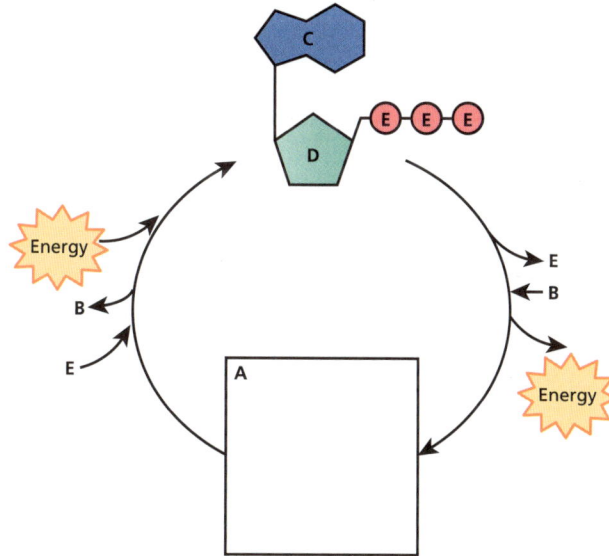

   a) Identify molecules A, B, C, D and E. [5]

   A = ...

   B = ...

   C = ...

   D = ...

   E = ...

   b) Draw the structure of molecule A in the box. [2]

   c) Write the names of **two** enzymes on the diagram to show which reaction each catalyses. [2]

   d) State the quantity of energy given out or taken in by each reaction. .... [1]

4. At any one moment, the human body contains about 50 g of ATP. However, in a day about 70 kg of ATP is made.

   a) Calculate the ratio of ATP made in a day : ATP content at any one moment. Give your answer in the form $x : 1$ [2]

   b) Explain why this ratio is so large. [4]

   ....

   ....

   ....

   ....

Water and inorganic ions

❓ Key words

❓ Key questions

What is a hydrogen bond?

What is cohesion?

Why is water a stable environment for animals to live in?

Water

Water is a major component of all living organisms. It has many important roles in organisms and they are linked to its properties. These properties are in turn related to the structure of the water molecule:

- A water molecule is made of two hydrogen atoms covalently bonded to one oxygen atom.
- The electrons shared by the atoms are pulled closer to the oxygen atom.
- This means that the oxygen has a slight negative charge (δ-) and the hydrogen is slightly positive (δ+).
- This means that the molecule is **polar** and the oxygen atom on one molecule is attracted to a hydrogen atom on another molecule.
- This attraction is called a **hydrogen bond**.

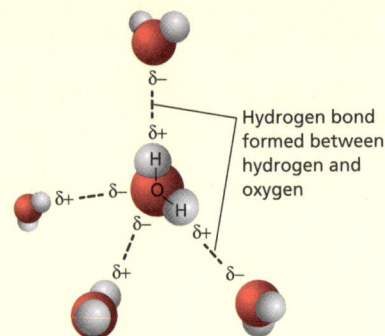

Hydrogen bond formed between hydrogen and oxygen

The water molecule's structure and the hydrogen bonds give water important properties:

- Due to the polarity of the molecules, water acts as a solvent, dissolving solutes such as inorganic ions, sugars and proteins. This means that many metabolic reactions can occur in water.
- The hydrogen bonds between the molecules mean that they are attracted to each other. This is called **cohesion** and it helps water to form long columns in the xylem of plants. It also provides surface tension, allowing some small animals to walk on the surface of ponds.
- Water has a relatively **high specific heat capacity,** so it takes a lot of energy to change its temperature. This provides a stable environment for aquatic organisms.
- The hydrogen bonds give water a large **latent heat of vaporisation**. This means it requires a lot of energy to evaporate water and so sweating can cool organisms.
- Water is an important reactant or product in many metabolic reactions. It is required in hydrolysis reactions to split molecules and is produced in condensation reactions. In photosynthesis, it provides hydrogen atoms to act as reducing power.

Inorganic ions

Inorganic ions are found in the cytoplasm and body fluids of organisms. Some are in high concentrations but others are in very low (trace) amounts. Some may be used to form important biological molecules.

The table shows the functions of some of these inorganic ions.

Inorganic ion	Symbol	Function
Hydrogen	H^+	- Determines pH: the more H^+ ions, the lower the pH - Provides the correct pH for enzymes in digestion - Is an important intermediary in respiration and photosynthesis
Iron	Fe^{2+}	Found in the haem group in the haemoglobin molecule
Sodium	Na^+	- Used in the co-transport of glucose and amino acids - Essential in the transmission of nerve impulses
Phosphate	PO_4^{3-}	A main component of DNA, ATP and phospholipids

✔ Summary

Water and inorganic ions

(1) The diagram shows three water molecules.

Draw **two** hydrogen bonds to link the three molecules. Use dotted lines. [2]

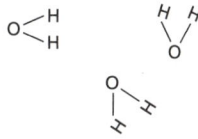

(2) Which of these statements about water are correct? Tick **one** box. [1]

It is produced when maltose is digested. ☐

It is required to convert dipeptides to amino acids. ☐

It is a product of hydrolysis of lipids. ☐

It is required to produce sucrose from glucose and fructose. ☐

(3) Draw lines to join each observation to the property of water that makes it possible. [3]

Observation	Property of water
Sea temperature shows less variation than air temperature	Cohesion
Glucose is carried dissolved in blood plasma	High specific heat capacity
Water can be drawn 100 m up the trunk of a tree	High latent heat
Evaporation of water from leaves cools plants	Polar molecule

(4) Bonding in water involves covalent bonds and hydrogen bonds.

Explain the difference between these bonds with reference to water. [4]

(5) The table shows the concentration of the main inorganic ions in human intracellular and extracellular fluid.

Inorganic ion	Concentration of ion (mM)	
	Intracellular fluid	Extracellular fluid
Na^+	10	145
K^+	140	5
H^+	7.0×10^{-5}	7.3×10^{-5}
Cl^-	10	110

a) Is the pH higher or lower inside cells compared to outside? Explain your answer. [2]

b) Glucose molecules can enter cells using a Na^+ cotransporter system.

Use the table to explain why this is possible. [2]

c) During a nerve impulse, the membrane of a neurone becomes permeable to sodium ions.

Use the data from the table to explain why the charge inside the cell becomes positive. [1]

Structure of eukaryotic cells

🔑 Key words

❓ Key questions

What structures are found in plant cells but **not** in animal cells?

Which cellular structures have a double membrane or envelope?

Why is rough ER described as 'rough'?

Eukaryotic cells

All living organisms are made of one or more cells. Cells can be **eukaryotic** or **prokaryotic**.

Animals, plants, fungi and protoctists all have eukaryotic cells and so are called eukaryotic.

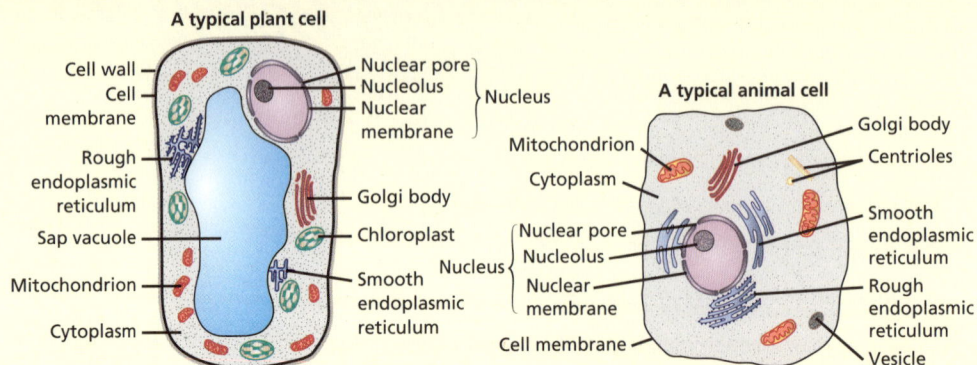

A typical plant cell

Cell wall — Cell membrane — Rough endoplasmic reticulum — Sap vacuole — Mitochondrion — Cytoplasm — Nuclear pore, Nucleolus, Nuclear membrane } Nucleus — Golgi body — Chloroplast — Smooth endoplasmic reticulum

A typical animal cell

Mitochondrion — Cytoplasm — Nuclear pore, Nucleolus, Nuclear membrane } Nucleus — Cell membrane — Golgi body — Centrioles — Smooth endoplasmic reticulum — Rough endoplasmic reticulum — Vesicle

Specialised structures found in eukaryotic cells

Feature	Structure	Function
Cell-surface membrane	Made of a phospholipid bilayer containing proteins (see page 38)	Determines which substances can enter and leave the cell
Nucleus	Surrounded by a membrane envelope and containing one or more nucleoli	Contains the genetic material, DNA; nucleoli produce RNA and ribosomes
Mitochondria	Surrounded by a membrane envelope, the inner membrane being folded into cristae; contain a fluid called the matrix	Site of aerobic respiration and ATP production (see page 86)
Chloroplasts	Surrounded by a membrane envelope and contain stacks of membranes arranged in grana; the fluid around the grana is called the stroma	Site of photosynthesis; the light-dependent reactions on the grana and the light-independent reactions in the stroma
Golgi apparatus	A series of flattened membrane sacs with vesicles; found near the nucleus	Proteins are packaged and modified for transport to the cell membrane
Lysosomes	Small membrane-bound vesicles containing hydrolytic enzymes	Used to digest unwanted organelles and molecules
Ribosomes	Small structures made of protein and RNA	Site of protein synthesis (see page 64)
Endoplasmic reticulum (ER)	A set of membrane channels that may have ribosomes on their surface	Rough ER with ribosomes is the site of protein synthesis; smooth ER without ribosomes transports lipids
Cell wall	Outer layer of plant cells, made of cellulose; made of chitin in fungi	Supports the cell and prevents bursting
Cell vacuole	Large membrane-bound structure in plant cells containing cell sap	Stores chemicals and when under pressure it supports the plant

✔ Summary

Structure of eukaryotic cells

(1) Which of these cellular structures have internal membranes? Tick **one** box. [1]

1. Nucleus   **2.** Chloroplasts   **3.** Nucleoli   **4.** Mitochondria

1, 2, 3 and 4 ☐   1 and 3 only ☐   2 and 4 only ☐   None of them ☐

(2) Which of these cellular structures contain nucleic acids? Tick **one** box. [1]

1. Ribosomes   **2.** Lysosomes   **3.** Golgi apparatus   **4.** Nucleoli

1, 2, 3 and 4 ☐   1, 2 and 3 only ☐   1 and 4 only ☐   2 and 4 only ☐

(3) Draw lines to join each sub-cellular structure with the correct feature of that structure. [3]

Structure	Feature
Lysosome	Transport of lipids
Golgi apparatus	Site of ribosome production
Smooth ER	Contains hydrolytic enzymes
Nucleoli	Modifies and packages proteins

(4) The diagram shows the parts of the cell responsible for the production and secretion of an enzyme from a cell.

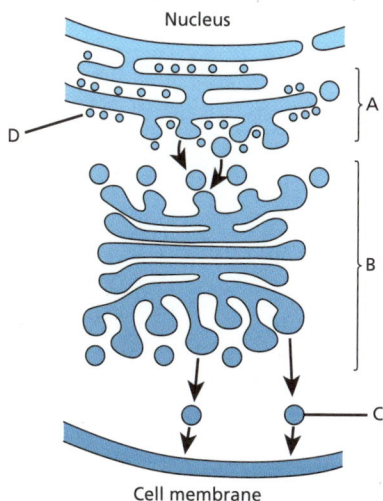

a) Name the structures A, B and D. [3]

A = ...

B = ...

D = ...

b) The cell is supplied with radioactive threonine, which is an amino acid.

The graph shows the level of radioactivity found in structures A, B and C over time.

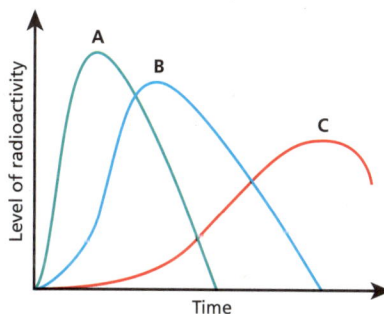

Explain the pattern of radioactivity found in structures A, B and C. [5]

....

....

....

....

....

Structure of prokaryotic cells and viruses

Key words

Key questions

What structures are found in eukaryotic cells but not in prokaryotic cells?

What is a plasmid?

Why are viruses not considered to be living?

Prokaryotic cells

Prokaryotic cells are different to eukaryotic cells because they do not have membrane bound organelles. They are also much smaller than eukaryotic cells. Scientists think that they existed early in the history of life and that eukaryotic cells evolved from prokaryotic cells.

Bacteria are prokaryotic. The diagram shows a typical bacterial cell.

- Like all prokaryotes, bacteria do not have a nucleus but have a circular chromosome, free in the cytoplasm.
- They have ribosomes but these are smaller than the ribosomes found in the cytoplasm of eukaryotes.
- There are no mitochondria, chloroplasts, lysosomes or endoplasmic reticulum.
- They do have a cell wall but, unlike plants or fungi, it is made from a glycoprotein called murein.
- Outside the cell wall is another layer called the capsule.
- They have small rings of DNA in the cytoplasm called **plasmids**.
- They may have one or more flagella that move them around.

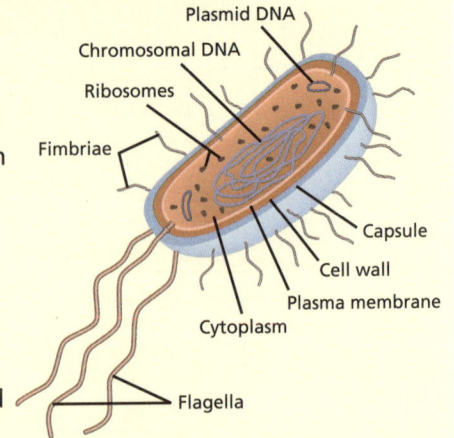

Labels: Plasmid DNA, Chromosomal DNA, Ribosomes, Fimbriae, Capsule, Cell wall, Plasma membrane, Cytoplasm, Flagella

Viruses

Viruses are not cells and so are not considered to be living organisms. They can only reproduce using living host cells. The viruses' genetic material instructs the host cell to make new viruses.

Viruses are much smaller than both eukaryotic and prokaryotic cells and have a wide variety of different shapes.

Viral shapes:
Polyhedral (adenovirus)
Spherical (influenza)
Helical (tobacco mosaic virus)
Complex (bacteriophage)

Viruses have certain features in common:
- They contain a strand of genetic material that is DNA in some viruses and RNA in others.
- The genetic material is surrounded by a protective protein coat called a **capsid**.
- Some viruses (enveloped viruses) have a lipid envelope that is made from the host cell's membrane. This may contain glycoproteins that attach the virus to the host cell.

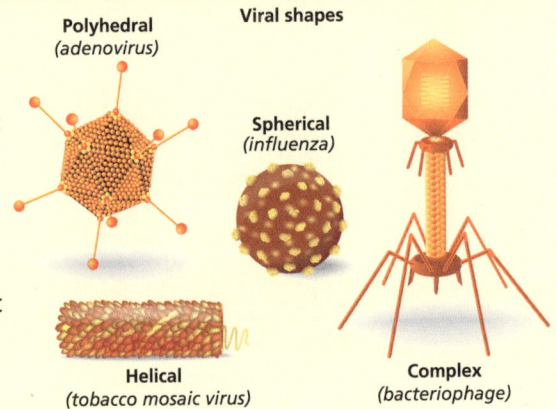

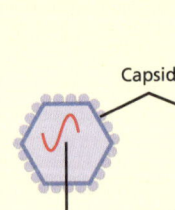

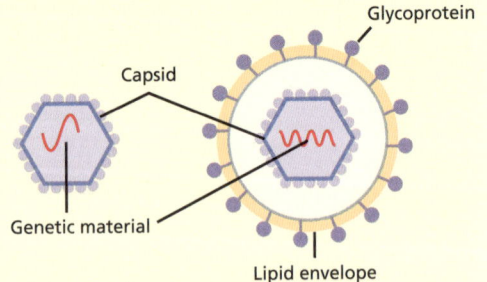

Non-enveloped virus
Enveloped virus
Labels: Glycoprotein, Capsid, Genetic material, Lipid envelope

Summary

Structure of prokaryotic cells and viruses

1. The table shows some features of prokaryotic cells, eukaryotic cells and viruses.

   Complete the table by putting a tick (✓) or a cross (✗) in each box. [5]

	Prokaryotic cells	Eukaryotic cells	Viruses
Contain ribosomes			
Contain cytoplasm			
Have a protein coat			
Contain plasmids			
Have a cell wall made of murein			

2. Which row in the table shows the correct nucleic acids found in prokaryotic cells and viruses? [1]

	Prokaryotic cells	Viruses
A	RNA only	DNA but never RNA
B	DNA and RNA	DNA or RNA
C	DNA or RNA	RNA but never DNA
D	DNA only	DNA or RNA

3. Explain the importance to viruses of the glycoproteins in their envelope. [2]

4. Scientists think that eukaryotic cells evolved from prokaryotic cells by a process called endosymbiosis.

   This process is shown in the diagram. It involves prokaryotic cells being engulfed by another cell to form the different organelles.

   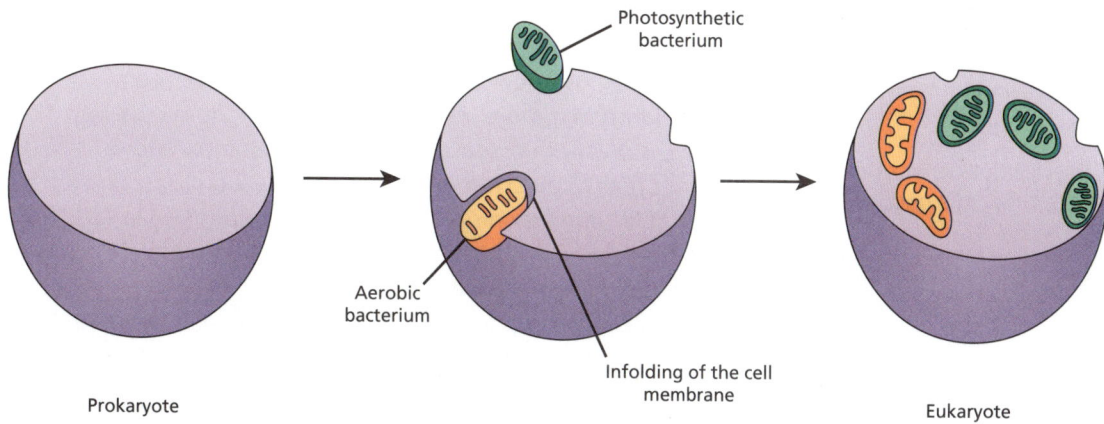

   a) Name the organelles formed by the bacteria shown in the diagram. [1]

   b) Use the endosymbiotic theory to explain why these two organelles have double membranes. [2]

   c) Explain why the ribosomes in eukaryotic cells are different to those in prokaryotic cells. [2]

   d) Explain why living organisms are not thought to have evolved from viruses. [2]

Methods used to study cells

Key words

Key questions

What does the term 'resolution' mean?

Microscopy

To view cells, a microscope is needed to provide sufficient resolution. **Resolution** is the ability to see two close objects as separate structures. The human eye can only differentiate between objects that are 0.1 mm apart. There are different microscope types.

Optical (light) microscopes pass visible light through a specimen and use lenses to focus the light and form an image.

- The magnification of the image is calculated by multiplying the magnification of the objective lens with that of the eyepiece lens.
- To allow different structures to stand out, the specimen is often treated with a stain.
- The magnification of any photograph or drawing of an image is calculated using the formula

$$\text{Magnification} = \frac{\text{size of image}}{\text{size of real object}}$$

- The limit of resolution using the light microscope is set by the wavelength of light and is about 0.2 µm. This allows cells to be seen and some internal detail, such as nuclei and the outline of mitochondria and chloroplasts.
- Thicker lenses can improve the magnification but not the resolution, so the limit of useful magnification is about x1500.

Light microscope

Image
Eyepiece lens
Objective lens
Specimen
Slide
Stage
Condenser lens

What limits the resolution of the light microscope?

Electron microscopes use a beam of electrons instead of visible light.

- Electron microscopes provide a much greater resolution than light microscopes as the wavelength of electrons is much smaller. This can give a resolution of 0.5 nm or better and allows much higher magnification to be usefully used and more detail to be seen.
- Specimen preparation is more involved than with light microscopy. The specimen must be dried, cut very thin and stained with substances such as heavy metals.
- Two different types of electron microscopes are used – transmission electron microscopes (TEM) and scanning electron microscopes (SEM). TEM fires electrons through the specimen and gives more detail. SEM bounces electrons off the specimen and gives a surface view.
- The complex specimen preparation takes some time and because the specimen must be dead and dried, this can introduce possible **artifacts** into the image.

What is the effect on a mixture of spinning it in a centrifuge?

Cell fractionation and ultracentrifugation

To investigate the activity of different structures in the cell, it is often necessary to separate the components. This involves two steps:

- **Cell fractionation** involves breaking up tissue and the cells to release the contents. This is done in an isotonic buffer solution to prevent changes owing to osmosis or pH. The extract produced is kept cold to stop any hydrolysis by enzymes.
- **Ultracentrifugation** uses a machine called a centrifuge to spin the extract at high speed. By varying the speed, components of different size will sink to the bottom of the tube (smaller components need higher speeds to sediment out).

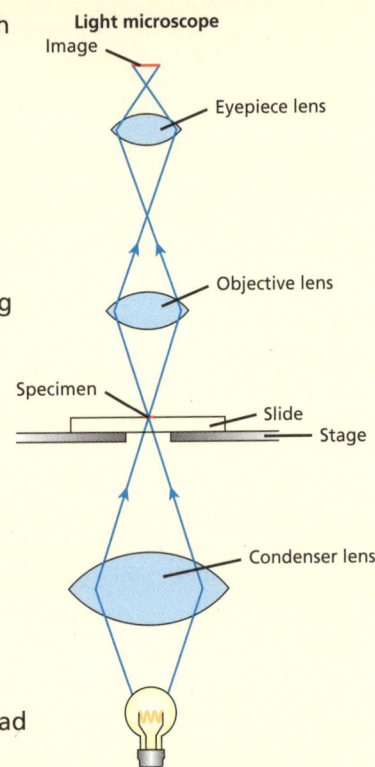

Summary

Methods used to study cells

1 Tick the box next to the correct meaning of resolution in microscopy. [1]

The diameter of the smallest object that can be seen. ☐

The wavelength of the radiation used to view the specimen. ☐

The power of the objective lens multiplied by the power of the eyepiece. ☐

The shortest distance apart that two objects can be seen separately. ☐

2 Which row in the table gives the correct values for resolution in **micrometres**? [1]

	Electron microscope	Light microscope	Naked eye
A	0.5	0.2	0.1
B	0.0005	0.2	100
C	0.5	200	100 000
D	5	2	1

3 Give **two** advantages of using an optical microscope rather than an electron microscope to view an object. [2]

....

....

4 The photograph shows part of a liver cell magnified 6000x.

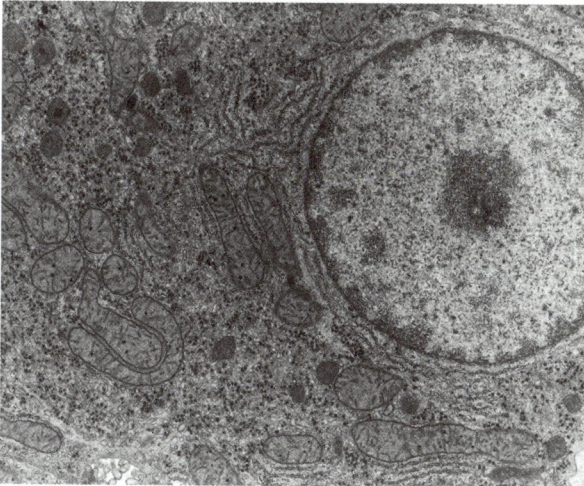

a) Name the type of microscope used to obtain this image. [1]

....

b) Give **three** different stages required in the preparation of this specimen. [3]

....

....

....

c) Calculate the actual diameter of the nucleus. Give your answer in micrometres. [3]

....

d) Scientists wish to isolate different organelles from liver tissue. They break open the cells in an ice-cold isotonic buffer solution. They then centrifuge the broken cells in an ultracentrifuge at three different speeds to obtain three pellets.

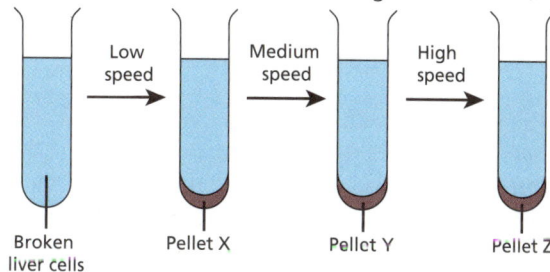

i) Name the process that breaks open the liver cells. [1]

....

ii) Explain why the isotonic buffer solution is ice-cold. [2]

....

....

iii) One of the pellets X, Y or Z contains nuclei, one contains ribosomes and one contains mitochondria. Identify the pellets. [3]

X = ...   Y = ...   Z = ...

Mitosis and the cell cycle

🔑 Key words

The cell cycle

In multicellular organisms, eukaryotic cells become specialised for specific functions. The cells become organised into tissues, tissues into organs, and organs into systems.

Within multicellular organisms, not all cells retain the ability to divide. Eukaryotic cells that do retain the ability to divide show a **cell cycle**.

The cell cycle is a set sequence of events that occur during the life of a cell:

❓ Key questions

What must happen to the DNA before mitosis can start?

- Most of the cycle is taken up by **interphase**. During G1 the cell grows, making new proteins and organelles. In S phase the DNA is replicated, and in G2 further growth occurs.
- After interphase, **mitosis** occurs. This divides the chromosomes into two sets and two new nuclei are formed.
- Cytokinesis (C) follows mitosis and divides the cytoplasm into two new cells.
- Various checks take place in the cell cycle to try to stop the production of cancerous cells.

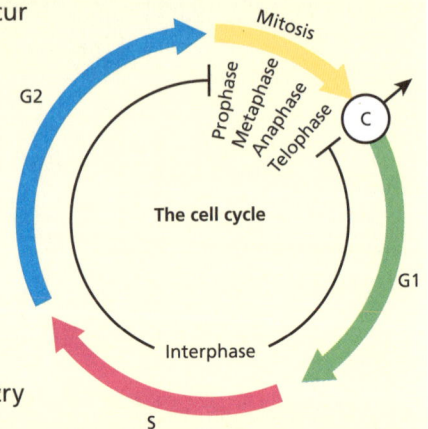

Mitosis

At the start of mitosis, each chromosome has already divided and the two copies are called **chromatids**. They are joined at the centre by a **centromere**. During the phases of mitosis, various events occur:

Why are chromatids pulled to opposite poles of the cell?

- In **prophase**, the chromosomes thicken and become visible. Centrioles move to opposite ends (poles) of the cell and produce spindle fibres made of microtubules.
- **Metaphase** follows and the chromosomes line up in the middle (equator) of the cell.
- During **anaphase**, the centromeres divide and the spindle fibres shorten, pulling the chromatids apart. They move to opposite poles of the cell.
- In **telophase**, the new chromosomes reach the poles and two new nuclear membranes form.

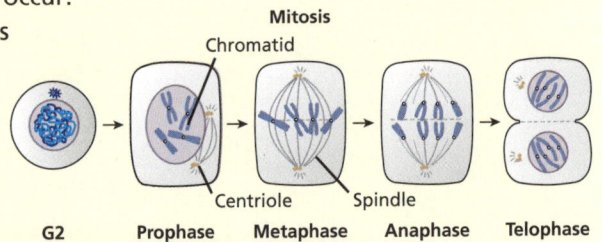

After telophase has finished, **cytokinesis** occurs. In animal cells, this involves the cell pinching in at the middle to form two new cells. In plant cells, a new cell wall grows along the equator to separate the cells.

What is cytokinesis?

Binary fission

Prokaryotic cells do not go through mitosis but divide by **binary fission**. The circular DNA and the plasmids divide and the DNA moves to opposite ends of the cell. The cytoplasm then divides to form two separate cells. This is a type of asexual reproduction. (Note that the plasmids might not be shared equally.)

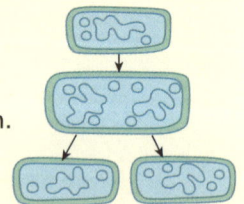

✔ Summary

Mitosis and the cell cycle

1. What happens in metaphase of mitosis? Tick **one** box. [1]

The DNA divides. ☐ The chromosomes gather at the equator of the cell. ☐

The spindle shortens. ☐ The chromosomes unwind and become thinner. ☐

2. Which row in the table gives the conditions in a human cell at the end of interphase? .... [1]

	Nuclear membrane present	Number of chromatids	Centrioles at either end of the cell
A	No	46	Yes
B	Yes	92	No
C	No	46	No
D	Yes	92	Yes

3. The drawings show the chromosomes in a cell at different stages of mitosis.

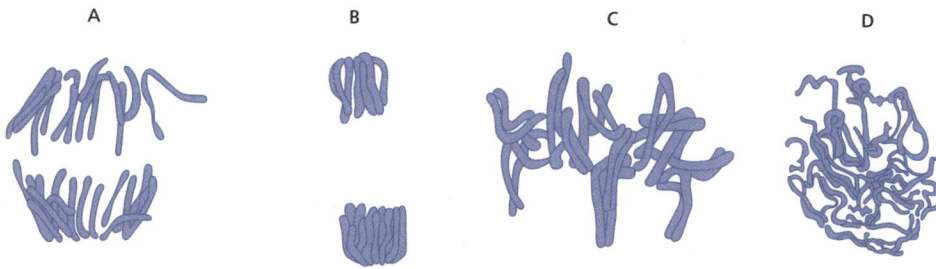

   a) Name the stage shown in each diagram. [4]

   A = ...   B = ...

   C = ...   D = ...

   b) Give a reason why chromosomes become visible during mitosis. [1]

   ....

   c) Between which two stages do the centromeres divide? [1]

   ....

4. The table shows some data for the number of cells at different stages of the cell cycle at any one time.

Stage	Number of cells
Interphase	510
Prophase	15
Metaphase	3
Anaphase	3
Telophase	2

   a) The cell cycle in this cell type takes 24 hours. The number of cells in each stage is proportional to the time spent in each stage.

   Calculate the length of time spent in anaphase. Give your answer in minutes, to the nearest whole number. [3]

   ....

   b) The mitotic index of a cell is calculated using this formula:

   mitotic index = number of cells in mitosis ÷ total number of cells

   i) Calculate the mitotic index for this cell type. [2]

   ....

   ii) The mitotic index is useful for detecting possible cancer.

   Give a reason why cancer cells have a high mitotic index. [1]

   ....

The cell membrane

Key words

Key questions

Why are the phospholipid tails on the inside of the bilayer?

Why is the cell membrane described as a fluid mosaic?

Why is cholesterol found in the centre of the membrane?

Phospholipid bilayers

The basic structure of all membranes in the cell is the same. This includes the cell-surface membrane and all the membranes around the cell organelles.

The main components are phospholipids (see page 12).

- Phospholipids have a hydrophilic, polar head due to the phosphate group and two hydrophobic, non-polar tails.
- The phospholipids form a bilayer.
- The polar heads are on the outside and the tails are in the centre of the bilayer.

A phospholipid

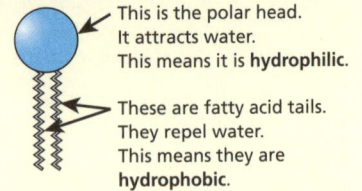

This is the polar head. It attracts water. This means it is **hydrophilic**.

These are fatty acid tails. They repel water. This means they are **hydrophobic**.

The fluid mosaic model

As well as containing a phospholipid bilayer, there are proteins embedded in the membrane. This is called the **fluid mosaic model**. This is shown in the diagram:

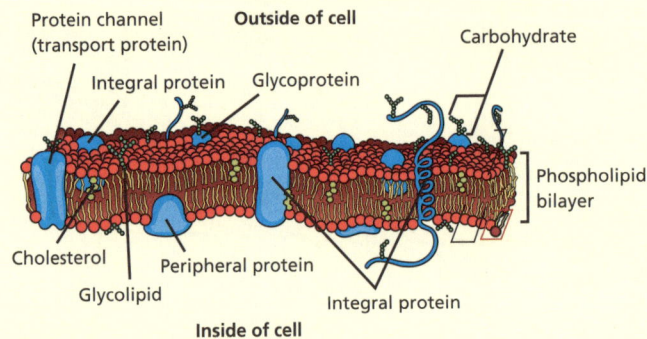

Protein channel (transport protein)
Outside of cell
Carbohydrate
Integral protein
Glycoprotein
Phospholipid bilayer
Cholesterol
Peripheral protein
Glycolipid
Integral protein
Inside of cell

These are the features of the fluid mosaic model:

- It is described as fluid because the proteins are free to move in the phospholipids. The phospholipids can also move.
- It is a mosaic because the proteins are scattered throughout the membrane.
- Integral proteins pass all the way through the membrane.
- Peripheral proteins are found on either surface of the membrane. On the outer surface of the cell membrane, these proteins often have carbohydrate chains attached.
- Cholesterol molecules are lipids and are found in-between the phospholipid tails. The cholesterol controls the fluidity of the membrane.

Functions of the cell membrane

The cell membrane has a number of different functions:

- Transport of substances into and out of the cell (see page 40).
- Identification of cells as self or foreign (see page 44).
- Attachment of enzymes.

Summary

RETRIEVE

The cell membrane

1. The diagram shows a molecule found in the cell membrane.

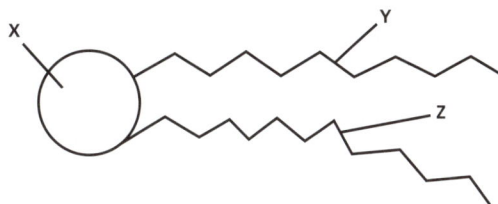

Which row of the table identifies a chemical group found in X, Y and Z? [1]

	X	Y	Z
A	Phosphate group	Unsaturated fatty acid	Saturated fatty acid
B	Peripheral protein	Saturated fatty acid	Glycerol
C	Phosphate group	Saturated fatty acid	Unsaturated fatty acid
D	Peripheral protein	Glycerol	Unsaturated fatty acid

2. The diagram shows a section through a cell membrane. [1]

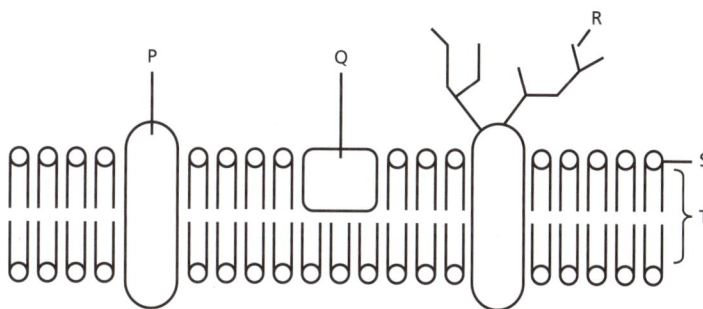

a) Name this model of the cell membrane. [1]

b) Give the letter from the diagram that fits each description. [4]

The area where cholesterol is found	
A carbohydrate molecule	
A peripheral protein	
A hydrophobic region	

3. Scientists investigate the phospholipids in the membrane of a cell.

They extract all the phospholipids and place them on a tray of water so they form a monolayer.

This is shown in the diagram.

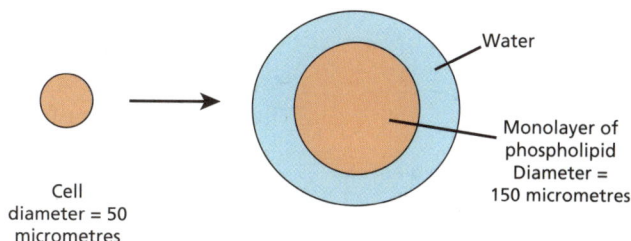

Cell diameter = 50 micrometres

Water

Monolayer of phospholipid Diameter = 150 micrometres

a) Calculate the surface area of the cell. Use the formula surface area of a sphere = $4\pi r^2$ and $\pi = 3.14$ [2]

b) Calculate the area of the phospholipid monolayer. [2]

c) Comment on the difference between the answers to part a) and part b). [2]

Cells 39

Transport across cell membranes

Methods of transport

Substances can travel across the cell membrane in different ways. The method depends on several factors such as molecule concentration, size, charge and lipid solubility.

Diffusion is the movement of a substance from an area where it is in high concentration to one in which it is in low concentration:

- No metabolic energy is needed for diffusion as the energy comes from the kinetic movement of the molecules.
- Lipid soluble/hydrophobic molecules, and small molecules such as gases and water, can diffuse through the membrane. Sugars and charged molecules cannot move by simple diffusion.
- The rate of diffusion depends on temperature, the concentration gradient and molecule size.

Facilitated diffusion, like simple diffusion, is a passive process moving molecules down a concentration gradient:

- It involves channel proteins or carrier proteins.
- These proteins allow molecules like charged ions to pass through the membrane.

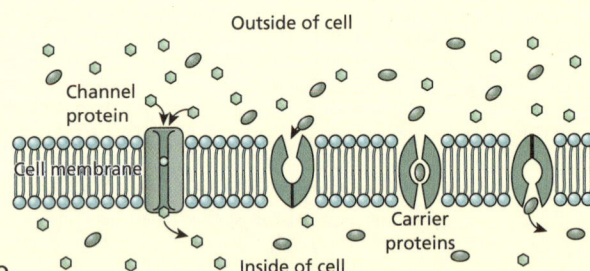

Osmosis is a special type of diffusion of water from a region of high water potential to a region of low water potential through a partially permeable membrane.

- Water potential is a measure of the concentration of free water molecules so a concentrated solution has a low or very negative water potential.
- In cells, the cell membrane is partially permeable.

Net movement from high water potential to low water potential

Active transport occurs when a molecule moves against the concentration gradient, from low to high concentration:

- It is an active process, so requires energy from respiration.
- The energy is in the form of ATP.
- It involves proteins in the membrane acting as carriers or pumps.

Cotransport is often called secondary active transport as it sets up a concentration gradient of one type of molecule and uses this gradient to move another molecule against its gradient. An example is the absorption of glucose in the ileum linked to Na$^+$ movement (see page 52).

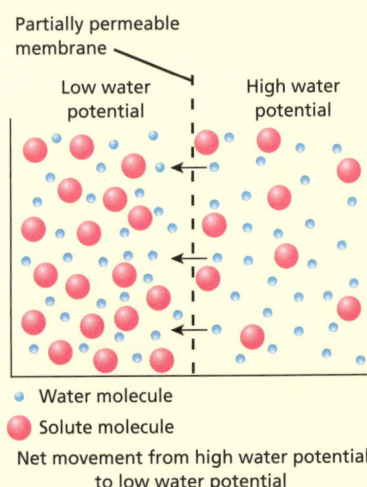

Adaptations for rapid transport

Cells can be adapted to increase the rate of transport across their membrane. For example:

- Folds or projections of the cell membrane to increase the surface area.
- An increase in the number of protein channels or carriers in the cell membrane.

Transport across cell membranes

(1) Which of these adaptations could increase the movement of a sugar into a plant cell by active transport? Tick **one** box. [1]

1. A decrease in the permeability of the cell wall

2. Infolding of the cell membrane

3. Increased concentration of sugars in the vacuole

4. More channel proteins in the cell membrane

1, 2, 3 and 4 ☐ 2, 3 and 4 only ☐ 2 and 4 only ☐ 1 and 4 only ☐

(2) Potato cylinders are placed in a 0.4 mol dm^{-3} sucrose solution. The cylinders do not change in mass.
The cylinders are then placed in a 0.8 mol dm^{-3} sucrose solution.

Which row of the table correctly describes the changes to the potato tissue? [1]

	Change in mass of the potato cylinder	Water potential in the potato cylinders
A	Increases	Increases
B	Decreases	Decreases
C	Decreases	Increases
D	Increases	Decreases

(3) The diagram shows the movement of glucose into a cell.

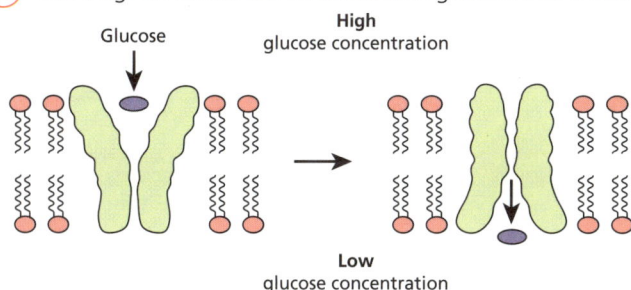

a) Name the type of transport that is occurring in the diagram. [1]

b) Explain why this is a passive process. [2]

(4) Scientists investigate the nitrate content in the roots of two barley plants, X and Y.

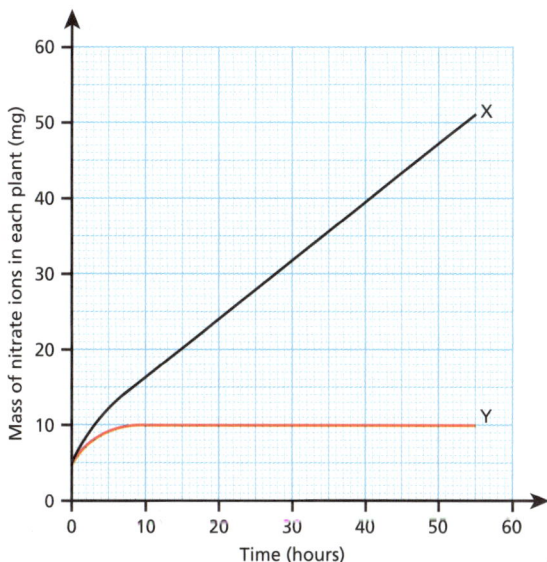

The plants were grown with their roots in a solution of minerals containing nitrate ions.

The solution for plant Y also contained a poison that inhibited respiration.

The mass of nitrate ions in the roots was measured for 60 hours.

The graph shows the scientists' results.

a) Calculate the percentage difference between the mass of nitrate in plant A compared with plant B after 60 minutes. [3]

b) Explain why there is a difference in the nitrate content of the two plants. [3]

c) Name the process causing the increase in the mass of nitrate in plant Y between 0 and 10 minutes. [1]

d) Explain why the increase in plant Y stops after 10 minutes. [2]

Antigens and phagocytes

Key words

Key questions

Where in the cell membrane are glycoproteins found?

Why do lysosomes bind with phagosomes?

What are macrophages?

Antigens

Antigens are chemical substances that can be recognised by the body's immune system. They are large organic molecules and are usually proteins or proteins with carbohydrates attached (glycoproteins). All body cells have antigens on the outer surface of their cell membranes.

Antigens can also be:
- found on the surface of micro-organisms and of structures such as pollen grains
- chemicals such as the toxins produced by bacteria
- chemicals taken into the body in food or injected by insects.

The body's immune system detects antigens to identify:
- pathogens, such as bacteria, fungi and viruses
- cells from other organisms of the same species
- abnormal body cells
- toxins.

Some pathogens show **antigenic variation**. Over time, the antigens on the surface are altered. This produces variants that the immune system finds more difficult to detect and destroy.

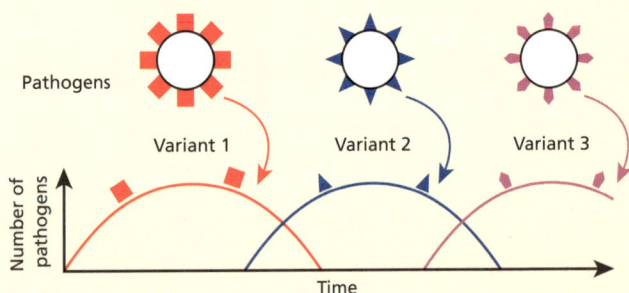

Phagocytes

If pathogens enter the body, the first line of defence by the immune system is called the non-specific immune response. This involves cells called phagocytes. Various types of white blood cells can act as phagocytes. They destroy pathogens by **phagocytosis**.

1. Binding sites on the surface of phagocytes adhere to antigens on the pathogens.
2. The phagocyte ingests the pathogen by surrounding it, forming a phagosome.
3. A lysosome fuses with the phagosome and releases enzymes, forming a phagolysosome.
4. The pathogen is digested by the enzymes.
5. The debris is then eliminated from the cell.

Phagocytosis often occurs in the blood. However, some white blood cells can squeeze out of capillaries into the tissues. They are then called **macrophages** and move around the tissues, engulfing pathogens.

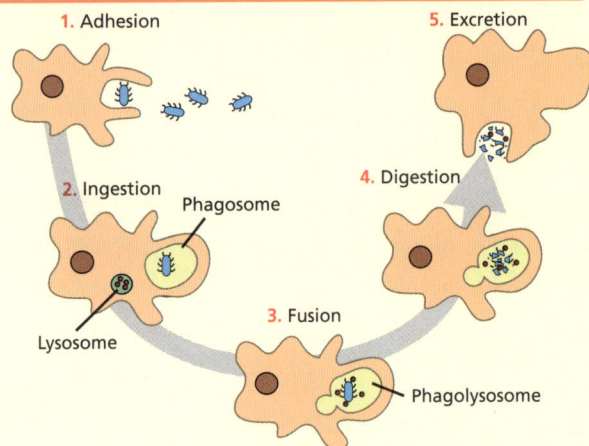

Summary

Antigens and phagocytes

1 The diagram shows part of the cell membrane.

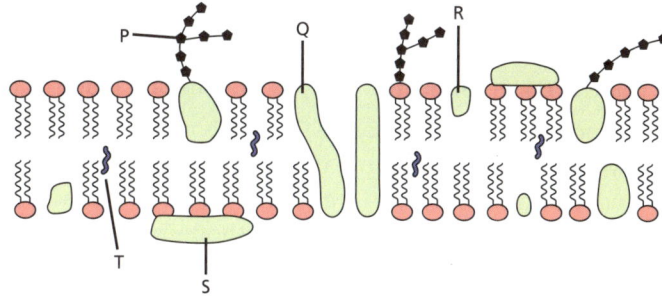

Which part of the cell membrane, P, Q, R, S or T, acts as antigens? .... [1]

2 Statements 1–4 give details about the adaptations of different types of pathogens.

1. Some pathogens replace their cell membrane glycoproteins with different ones every seven days.

2. Some pathogens cover their body surface with the host's proteins.

3. Some pathogens spend time inside the host's liver or red blood cells.

4. Some pathogens produce chemicals that inactivate white blood cells.

a) Which statement is an example of antigenic variation? .... [1]

b) For each statement, explain how the adaptation helps protect the pathogen. [4]

Statement 1 ....

....

Statement 2 ....

....

Statement 3 ....

....

Statement 4 ....

3 Scientists investigate the rate of phagocytosis by macrophages.

They use two strains of macrophages: a normal strain and a strain that was lacking a type of unsaturated phospholipid in the cell membrane. The unsaturated fatty acid determines the fluidity of the cell membrane.

The scientists mixed the macrophages with microscopic fluorescent beads and measured the rates of phagocytosis.

The graph shows the scientists' results.

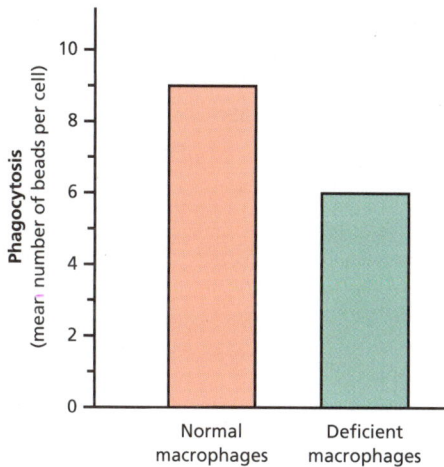

a) Give a reason why the beads were made fluorescent. [1]

....

....

b) Calculate the percentage decrease in the deficient macrophages compared with the number of beads in the normal macrophages. [3]

c) Explain why there are less beads in the deficient macrophages. [3]

....

....

The specific immune response

Key words

Key questions

What are antigen-presenting cells?

Why do antigens only bind with a specific B cell?

What is agglutination?

T lymphocytes

If pathogens are not destroyed by the non-specific actions of the phagocytes, a more specific immune response occurs, which is the result of other white blood cells called **lymphocytes**. All lymphocytes are made in the bone marrow.

The action of **T lymphocytes** (T cells) is called the **cellular response**. Once produced, T lymphocytes spend some time in the thymus gland. T cells have an antigen-binding site on their surface. Other cells, such as macrophages, destroy pathogens and present the antigens of the pathogens on their surface. They are then known as **antigen-presenting cells**. The binding sites of T cells bind with the antigens and the T cells are activated. There are different types of T cells:

- **Helper T cells** (T_H cells) divide when activated by antigens. They release chemicals called cytokines, which attract phagocytes and other types of T cells to the area and activate them.
- **Cytotoxic T cells** (T_C cells) detect antigens on infected, cancerous or damaged cells. They then destroy the cells.

B lymphocytes

B lymphocytes (B cells) are responsible for the humoral immune response. Their main function is to produce **antibodies**:

- Each B cell has a receptor for one specific antigen. When this antigen binds, it is called **clonal selection**.
- The B cell is then activated by a helper T cell and then divides to form a clone of **plasma cells**. This is called **clonal proliferation**.
- The plasma cells then produce specific antibody molecules.

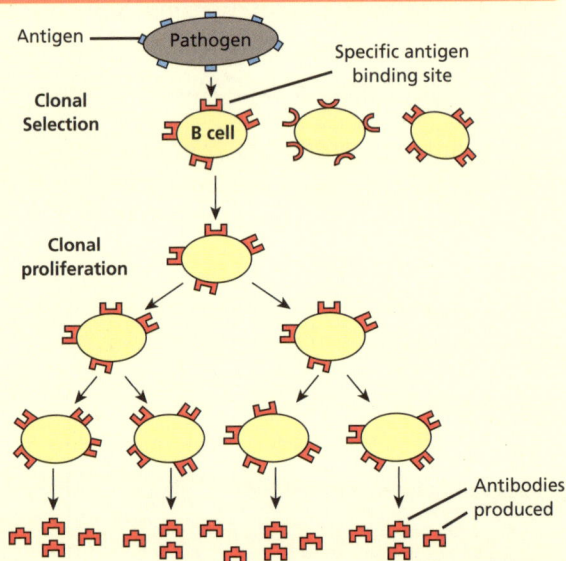

Antigen — Pathogen — Specific antigen binding site

Clonal Selection — B cell

Clonal proliferation

Antibodies produced

Antibodies

Antibodies are specialised protein molecules sometimes called **immunoglobulins**.

They bind with specific antigens to form an **antigen-antibody complex**. As each antibody has two binding sites, the antigens clump together. This is called **agglutination**. The pathogens can then be destroyed by phagocytes via phagocytosis. The structure of an antibody molecule is adapted to its function.

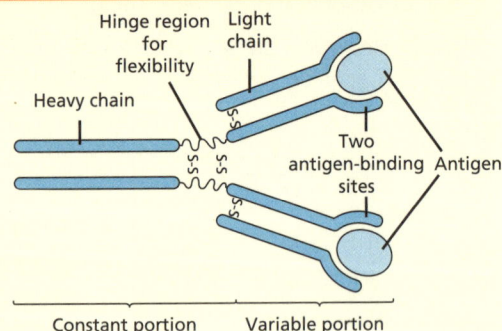

Hinge region for flexibility — Light chain

Heavy chain

Two antigen-binding sites — Antigen

Constant portion — Variable portion

Summary

The specific immune response

(1) B cells are produced when stem cells divide.

Where in the human body does this occur? Tick **one** box. [1]

Bone marrow ☐   Liver ☐

Lymph glands ☐   Thymus gland ☐

(2) Tick **one** box to show which statements about T cells are correct. [1]

1. Can stimulate cells to carry out phagocytosis

2. Can destroy cancer cells

3. Can activate B cells

4. Are responsible for the cellular response

1, 2, 3 and 4 ☐   1, 2 and 3 only ☐

3 and 4 only ☐   2 and 4 only ☐

(3) Antibodies are an important part of the humeral response.

a) State the number of polypeptide chains in an antibody molecule. .... [1]

b) State the number of antigen-binding sites in an antibody molecule. .... [1]

c) Explain the importance of the number of binding sites. [2]

....

....

d) Before a person is born, negative selection of B cells occurs. This kills all B cells that can bind to the person's antigens. Explain why negative selection is so important. [2]

....

....

(4) The diagram shows processes involved in the humeral response.

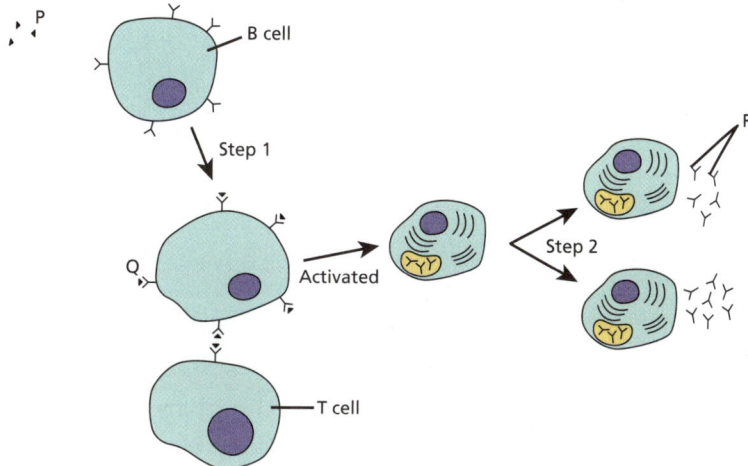

a) Name the structures labelled P, Q and R. [3]

P = ...

Q = ...

R = ...

b) Name step 1 shown in the diagram. .... [1]

c) Name the type of cell division shown in step 2. .... [1]

d) Explain why it is important that it is this type of cell division. [2]

....

....

e) Name the type of T cell shown in the diagram. .... [1]

Immunity, HIV and monoclonal antibodies

🔑 Key words

❓ Key questions

What is the difference between active and passive immunity?

Immunity

When a pathogen has triggered clonal selection of B cells, the plasma cells that are produced make antibodies. This is called the **primary immune response**. However, some of the B cells develop into memory cells rather than plasma cells. The memory cells are responsible for the **secondary immune response**:

- The memory cells can survive in the blood for many years.
- If the pathogen reinvades the body, the memory cells detect the antigens.
- The memory cells form plasma cells, which produce large numbers of antibodies very quickly and the pathogen is killed before the person becomes ill.

This type of **immunity** is called **active immunity** because the person's immune system produces the antibodies. In **passive immunity**, a person receives antibodies that have been made by another person or animal. An example is a baby receiving antibodies in their mother's breast milk.

Vaccination involves a person receiving medication to stimulate a primary immune response. The vaccination may contain dead or weakened pathogens, or just the surface antigens. Memory cells are therefore produced and if the person is infected with the harmful pathogen, they can mount a secondary response and avoid illness.

- If a certain percentage of the population is vaccinated, it prevents anyone from developing the disease. This is **herd immunity** and is about 90% of the population.
- The pathogen can show antigenic variation (see page 42), which means that a new vaccine is needed to protect against the new strain of pathogen.
- A small number of people may have harmful reactions to a vaccination. It is important to consider the benefits of the vaccination compared to these risks. If some people opt not to be vaccinated, herd immunity can be jeopardised.

Why does HIV contain the enzyme reverse transcriptase?

HIV

HIV is the virus that causes AIDS (acquired immune deficiency syndrome). The virus replicates by fusing with the surface of T helper cells. The RNA then enters the cell and is converted to DNA by reverse transcriptase. The DNA uses the host cell to make new copies of the virus. This destroys the host cell. When T helper cell numbers fall, an infected person cannot fight off infections and some types of cancer.

HIV (Human immunodeficiency virus)
- Viral envelope
- Capsid
- Reverse transcriptase
- RNA (two identical strands)
- Glycoprotein

HIV cannot be treated with antibiotics as it is a virus. Infected people are given antiviral drugs, which stop the virus replicating.

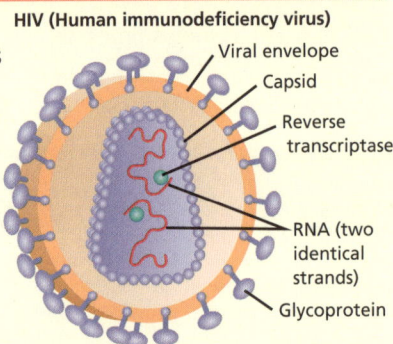

What does the ELISA test indicate?

Monoclonal antibodies

Monoclonal antibodies are made in the lab by cloning a specific immune cell. This produces large quantities of identical antibodies, which can then be used in:

- targeted medication by attaching a therapeutic drug to an antibody. The drug will then be delivered to a specific cell type that the antibody will bind to. This could be a cancer cell if a specific antigen can be found on the cell.
- medical diagnosis by identifying specific proteins or antigens. This could be a hormone in pregnancy tests or pathogens in Covid lateral flow tests. In the ELISA test, the blood is tested not for the pathogen but for antibodies that have been produced in response to the pathogen. But the ELISA test does not tell if the patient is still infected or is infectious.

✓ Summary

Immunity, HIV and monoclonal antibodies

1 Which of these is an example of passive immunity? Tick **one** box. [1]

Rapid production of antibodies in response to a pathogen ☐

Immunity after receiving an injection containing dead bacteria ☐

Production of memory cells ☐

A fetus gaining antibodies across the placenta ☐

2 The diagram shows a test for a pathogen using monoclonal antibodies.

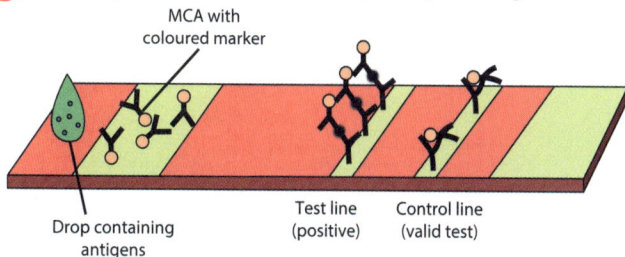

MCA with coloured marker

Test line (positive)   Control line (valid test)

Drop containing antigens

Which is a correct feature of the test? Tick **one** box. [1]

Antibodies from the pathogen are carried across the plate. ☐

Antibodies bind to antibodies on the test and control line. ☐

Antigens bind to antibodies on the test line. ☐

Coloured markers bind to antibodies on the test and control lines. ☐

3 The graph shows the concentration of antibodies in the blood after a person is given two doses of a pathogen.

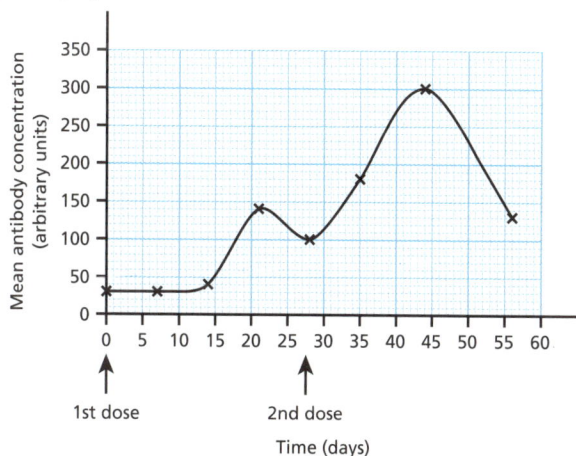

Mean antibody concentration (arbitrary units)

1st dose   2nd dose

Time (days)

a) The antibody level rises to a higher maximum after the second dose.

Calculate the ratio of the maximum after the second dose compared to the maximum after the first dose. [2]

b) Give **one** other difference between the response to the second dose and the response to the first dose. [1]

c) Explain the difference between the responses to the two doses. [2]

4 The graph shows the number of T helper cells and HIV viruses in the blood of a person after they have been infected by HIV. The person has not received any treatment.

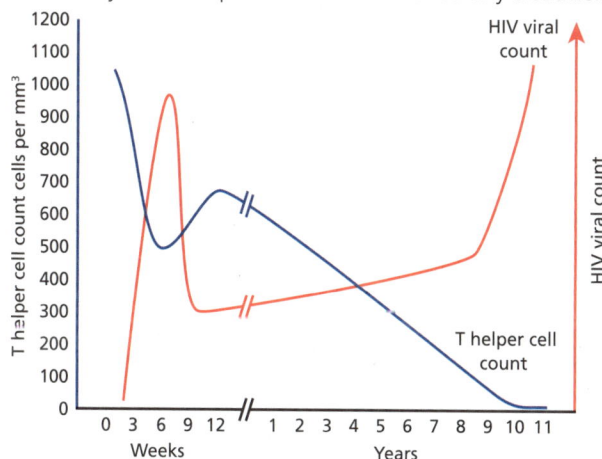

T helper cell count cells per mm³

HIV viral count

HIV viral count

T helper cell count

Weeks   Years

a) Describe and explain the relationship between the T helper cell count and the HIV count. [2]

b) A person does not usually show symptoms until their T helper cell count is below 200 per mm³.

Explain what implications this has for diagnosing HIV. [2]

Principles of gas exchange

Key words

Key questions

Why is the SA : Vol of an organism so important?

What is the function of ventilation movements in insects?

What are xerophytes?

Surface area to volume ratio

As organisms become larger, their volume increases. This means they have to exchange more substances with the environment. The rate at which they can exchange substances depends on the surface area. It is therefore important to know the **surface area to volume ratio** (SA : Vol).

- The larger the organism, the smaller the SA : Vol.
- This means it is harder for large organisms to obtain enough resources or remove waste fast enough.
- Large organisms therefore have adaptations to increase the rate of exchange.

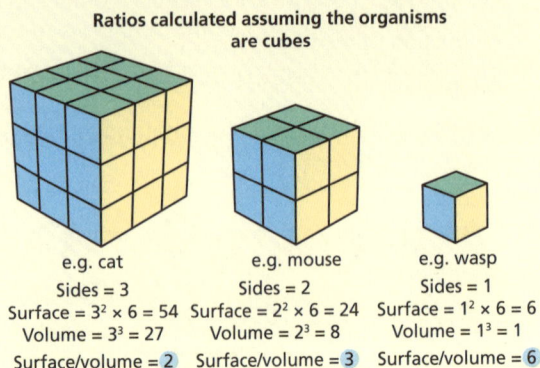

Ratios calculated assuming the organisms are cubes

e.g. cat
Sides = 3
Surface = $3^2 \times 6 = 54$
Volume = $3^3 = 27$
Surface/volume = 2

e.g. mouse
Sides = 2
Surface = $2^2 \times 6 = 24$
Volume = $2^3 = 8$
Surface/volume = 3

e.g. wasp
Sides = 1
Surface = $1^2 \times 6 = 6$
Volume = $1^3 = 1$
Surface/volume = 6

Gas exchange surfaces

Single-celled organisms do not need specific adaptations for gas exchange. They have a large SA : Vol, so rate of diffusion across the whole cell membrane is sufficient.

Insects have a tracheal system for gas exchange:
- Air enters an insect through openings called spiracles on the outside of its body.
- The spiracles are connected to tubes called tracheae.
- The tracheae run deep into the insect's body, ending as fluid filled tracheoles.

Adaptations to achieve sufficient exchange include:
- tracheoles connecting directly to cells, ensuring that all cells can exchange gases
- fluid being removed from the tracheoles during exercise
- ventilation movements in larger insects to move air in and out of the trachea.

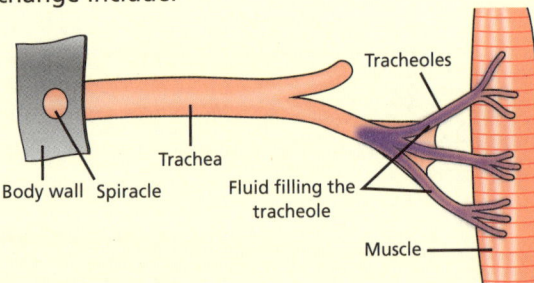

Tracheoles

Trachea

Body wall Spiracle

Fluid filling the tracheole

Muscle

Fish use gills, which have a high SA : Vol, for gas exchange:
- Gills have many filaments or lamellae. Each lamella has secondary lamellae that further increase the surface area.
- To maximise the diffusion gradient, a counter-current flow mechanism is used so that oxygen-rich water flows past a fish gill in the opposite direction to the blood.

Dicotyledonous plants have leaves that have adaptations for gas exchange, including:
- stomata, that are opened or closed by guard cells
- air spaces in the spongy mesophyll for diffusion.

Insects, and plants that live in areas of low water availability (xerophytes), have to balance the need to exchange gases with reducing water loss:
- Insects have a waterproof cuticle restricting exchange to the spiracles.
- Xerophytes have leaf adaptations such as thick cuticles, hairs, sunken stomata.

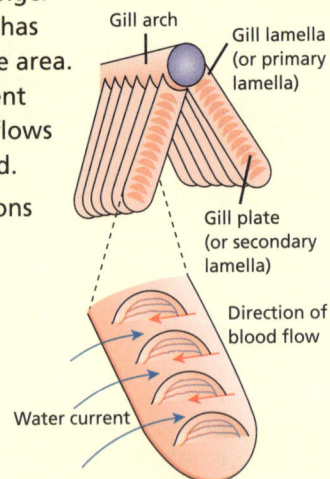

Gill arch

Gill lamella (or primary lamella)

Gill plate (or secondary lamella)

Direction of blood flow

Water current

Summary

Principles of gas exchange

(1) Which of these adaptations increase gas exchange in fish? Tick **one** box. [1]

 1. A parallel flow system between blood and water

 2. A low concentration of oxygen in the water

 3. The presence of lamellae in the gills

 1, 2 and 3 ☐ 1 and 3 only ☐ 1 and 2 only ☐ 3 only ☐

(2) The diagram shows representations of three different organisms, **A**, **B** and **C**. **A** is a sphere, **B** is a cube and **C** is a rectangular cuboid.

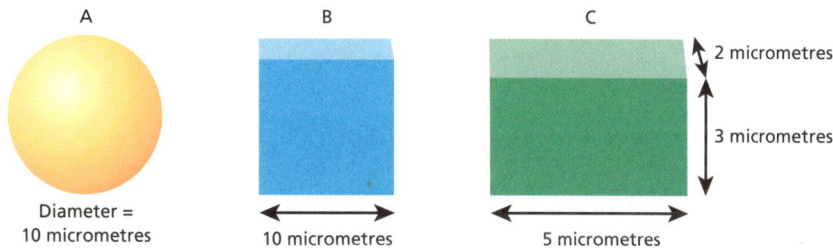

A

B

C

2 micrometres

3 micrometres

Diameter =
10 micrometres

10 micrometres

5 micrometres

(Not drawn to scale)

a) Complete this table for the three organisms. Use: volume of a sphere $= \frac{4}{3}\pi r^3$ and $\pi = 3.14$ [5]

Organism	Surface area (micrometres^2)	Volume (micrometres^3)	SA : Vol
A	314		
B		1000	0.6 : 1
C	62		

b) Discuss what the results in the table show about the rate of gaseous exchange in these three organisms. [3]

....

....

....

(3) The graph shows the concentration of oxygen in the tracheoles of an insect when it is sitting on a branch.

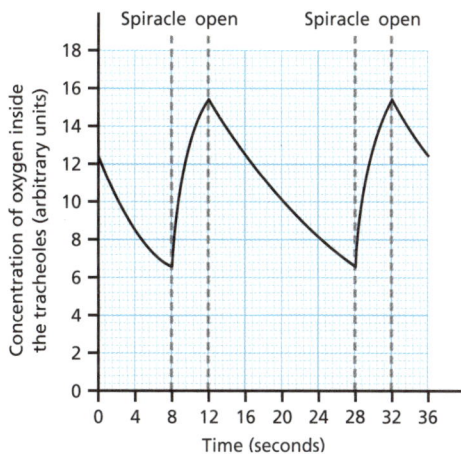

Spiracle open   Spiracle open

Concentration of oxygen inside the tracheoles (arbitrary units)

18
16
14
12
10
8
6
4
2
0

0 4 8 12 16 20 24 28 32 36

Time (seconds)

a) Explain why insects open and close their spiracles rather than having them open all the time. [2]

....

....

....

b) Calculate how many times a minute the spiracles open. [2]

....

c) Calculate the percentage of time that the spiracles are open. [2]

....

d) Explain why this percentage would change if the insect started flying. [3]

....

....

Gas exchange in humans

[QR code]

🔑 Key words

❓ Key questions

Why do humans need a specialised gas exchange system?

The two sets of intercostal muscles are antagonistic. What does this mean?

What are risk factors?

The human gas exchange system

Gas exchange occurs in the lungs, which contain millions of small sacs called **alveoli**. A series of tubes connect the alveoli to the outside.

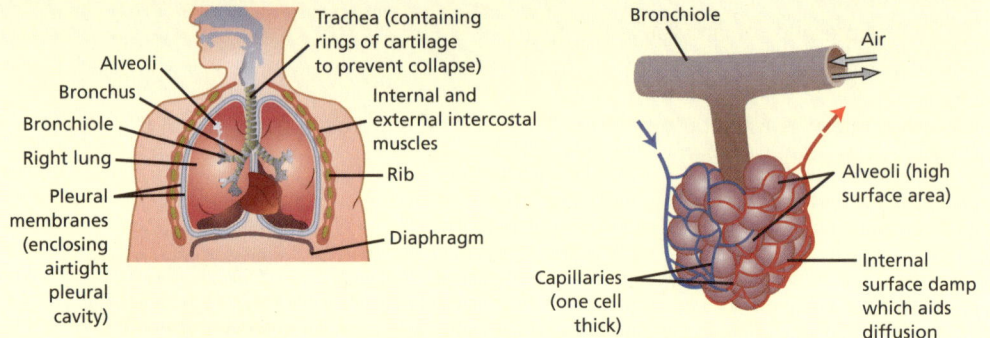

These adaptations increase the rate of gas exchange:
- Millions of alveoli increase the total surface area.
- There is a rich blood supply to maintain the diffusion gradient.
- The epithelial walls of the alveoli and the endothelial walls of the capillaries are each one cell thick and flattened, so reducing the diffusion pathway.
- There is a method of ventilation to increase the diffusion gradient.

Ventilation

Ventilation involves muscular contractions that move air in and out of the tubes leading to the alveoli.

Inhalation	Exhalation
When air is inhaled (breathed in), this occurs: 1. The external intercostal muscles contract causing the ribs to move upwards and outwards. 2. The diaphragm contracts, causing it to flatten. 3. The volume inside the pleural cavity increases and, because it is airtight, the pressure decreases. 4. The air outside the lungs is at a higher pressure so flows down a pressure gradient from high to low pressure.	Exhalation (breathing out) is the reverse of inhalation: 1. External intercostal muscles relax and internal intercostals contract, causing the ribs to move down and inwards. 2. The diaphragm relaxes and domes upwards. 3. The volume of the pleural cavity has decreased so the pressure is higher than the outside air. 4. Air inside the lungs passes out down a pressure gradient, high to low air pressure.

The pattern of breathing can be investigated using a device called a **spirometer**. This produces a trace of the breathing pattern that can be used to calculate breathing rate and volume of air exchanged.

Various factors can alter the efficiency of breathing:
- Tobacco smoking can increase risk of lung diseases, such as cancer and bronchitis.
- Pollution can increase the risk of bronchitis and asthma.
- Genetic factors, for example alleles causing disorders such as cystic fibrosis.

✔ Summary

Gas exchange in humans

(1) In 1950, a scientist called Richard Doll investigated the cause of death of a large number of patients in a London hospital. The table shows his observations.

	Number of deaths in a year per thousand patients	
	Non-smokers	**Heavy smokers**
Deaths due to lung cancer	0.0	1.1
Deaths by all causes	13.6	16.3

Which of these statements can be concluded from these results? Tick **one** box.   [1]

1. Smoking causes lung cancer.

2. It is impossible for non-smokers to get cancer.

3. Heavy smokers are more likely than non-smokers to die from diseases other than lung cancer.

4. Smoking increases the risk of dying from lung cancer.

1, 2, 3 and 4 ☐   1, 3 and 4 only ☐   3 and 4 only ☐   2 and 3 only ☐

(2) The diagram shows an alveolus surrounded by several capillaries. The magnification is ×3500.

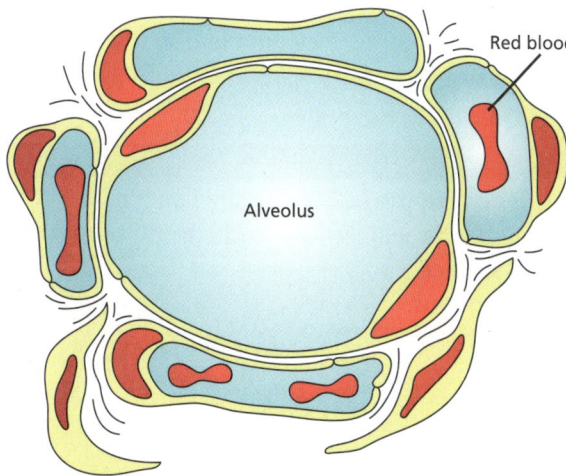

Red blood cell

Alveolus

a) Oxygen molecules diffuse from the alveolus into the red blood cell labelled on the diagram.

Give the number of membranes the oxygen molecules need to diffuse through.   [1]

b) Calculate the minimum distance the oxygen molecules have to diffuse to make this journey.

Give your answer in micrometres.   [2]

c) Give the features of the alveolus and the capillary that ensure that this distance is small.   [2]

(3) The graph shows a breathing trace for a person at rest. When the trace goes up, the person is breathing out and when it goes down the person is breathing in.

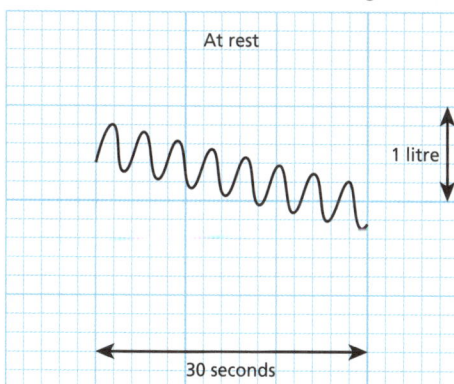

At rest

1 litre

30 seconds

a) Draw an X at one place on the graph where the external intercostal muscles are contracting.   [1]

b) Deduce the breathing rate in breaths per minute at rest.   [1]

c) The pulmonary ventilation rate (PVR) is given by this equation:

PVR = breathing rate × volume of one breath

Calculate the PVR at rest.   [2]

d) Explain why the PVR would be different if the person was exercising.   [3]

Digestion and absorption

🔒 **Key words**

❓ Key questions

What is the difference between an endopeptidase and an exopeptidase?

Why is lipid digestion faster with the presence of bile salts?

Why can monoglycerides diffuse through the cell membrane?

Digestion

Digestion involves the hydrolysis of large molecules to smaller molecules so that they can be absorbed across cell membranes and pass into the bloodstream. Digestion occurs in the alimentary canal and is catalysed by various enzymes.

Carbohydrates need to be hydrolysed to monosaccharides to be absorbed:
- **Amylase** enzymes digest starch into the disaccharide maltose. This occurs in the mouth (by salivary amylase) and in the duodenum (by pancreatic amylase).
- Disaccharides such as maltose are then digested by disaccharidase enzymes on the surface of the cells lining the ileum (in the case of maltose, the enzyme is maltase).

Proteins need to be hydrolysed to amino acids:
- Pepsin in the stomach and trypsin in the duodenum are **endopeptidases**, breaking peptide bonds in the middle of polypeptide molecules.
- The resulting shorter peptides are hydrolysed by **exopeptidases** to dipeptides.
- The dipeptides are digested to amino acids by dipeptidases on cells lining the ileum.

Lipids in lipid droplets need to be digested to release fatty acids:
- The lipid droplets are first **emulsified** in the duodenum into smaller droplets by the action of **bile salts** from the liver.
- The smaller droplets have a larger combined surface area and so the lipid molecules can be more rapidly hydrolysed to fatty acids and monoglycerides by lipase molecules from the pancreas.

Absorption

The products of digestion are absorbed into the cells lining the ileum by a combination of different mechanisms. Folds, villi and microvilli (brush border) on the surface of the ileum increase the surface area for absorption. The molecules then pass into the bloodstream.

Amino acids and monosaccharides are absorbed by co-transport mechanisms (see page 40).

The glucose or amino acid molecules attach to a binding site on the carrier molecule. When Na^+ binds, it induces a conformational change in the shape of the carrier protein, which draws the Na^+ ions and the glucose or amino acid molecules through the protein, in the same direction.

After digestion of triglycerides by lipase, the resulting monoglycerides and fatty acids form structures called **micelles**.

When close to the cells of the ileum, the monoglycerides and fatty acids move out of the micelles and diffuse into the cells.

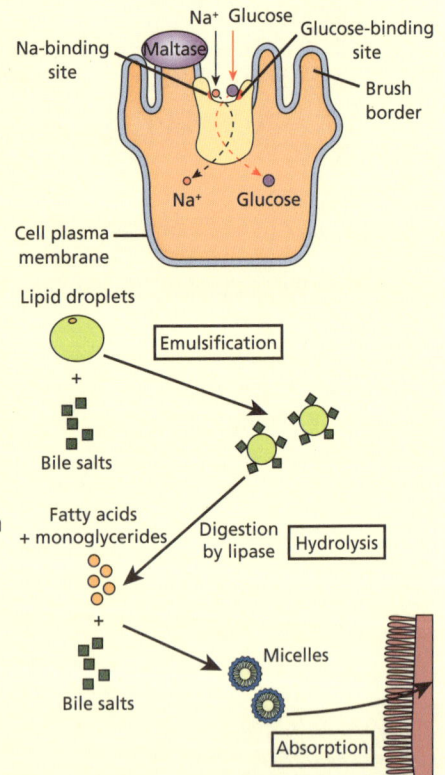

✅ **Summary**

Digestion and absorption

1 Which of these statements about carbohydrate digestion and absorption is correct? Tick **one** box. [1]

1. Amylase hydrolyses starch to glucose.
2. Salivary amylase and pancreatic amylase catalyse different reactions.
3. Maltose digests maltase in the duodenum.
4. Maltose is absorbed into cells in the ileum.

1, 2, 3 and 4 ☐   1 and 4 only ☐   2 and 3 only ☐   None of the above ☐

2 Glucose and monoglycerides are both absorbed by cells in the ileum.

a) Name the processes that are responsible for the absorption of these two substances. [1]

glucose = ...   monoglycerides = ...

b) Explain why different processes are needed. [2]

....

....

c) Explain why the location of maltase in the ileum increases the rate of glucose absorption. [2]

....

3 A student uses five test tubes and pours milk, sodium carbonate solution and an indicator into each tube. The indicator turns pink if the pH falls below 8.5. The student then adds other substances to the tubes and times how long it takes for the contents to turn pink. The table shows the other substances in each tube and the time taken to change colour.

Test tube	Other substances added to the tube	Time taken to turn pink (minutes)
1	1 cm^3 lipase solution + 1 cm^3 water	10
2	1 cm^3 bile salts solution + 1 cm^3 water	> 25
3	1 cm^3 lipase solution + 1 cm^3 bile salts solution	5
4	1 cm^3 boiled lipase solution + 1 cm^3 bile salts solution	> 25
5	1 cm^3 lipase solution + 1 cm^3 boiled bile salts solution	5

a) Explain why the indicator turned pink in tubes 1, 3 and 5. [2]

....

b) Explain why the colour change in tubes 3 and 5 occurred faster than in tube 1. [3]

....

....

c) Explain the difference in the time taken to change colour in tube 4 compared to tube 5. [3]

....

....

4 Pepsin is an endopeptidase enzyme active in the stomach. It is released by the cells lining the stomach in an inactive form and is then activated by the acid in the stomach.

a) Suggest why pepsin is released in an inactive form. [2]

....

....

b) In the small intestine, exopeptidases further digest the products of pepsin digestion.

Explain why it is more efficient to digest proteins using endopeptidases and exopeptidases, rather than just exopeptidases. [2]

....

....

c) Name the enzyme that acts on the product of exopeptidase digestion. .... [1]

Haemoglobin and oxygen carriage

Why is it important that haemoglobin binds reversibly with oxygen?

What shape is the oxyhaemoglobin dissociation curve?

What is the Bohr shift?

Structure of haemoglobin

Haemoglobins are a group of globular protein molecules that transport oxygen in the red blood cells of all vertebrates and in the tissues of some invertebrates.

Many animals are adapted to their environment by possessing different haemoglobins that have slightly different properties.

Human haemoglobin:

- is a quaternary globular protein, made up of four subunits (two α units and two β units)
- has four non-protein **haem** groups tightly associated to the protein.

Each haem group has an iron atom at the centre. An oxygen molecule can bind reversibly with each haem group.

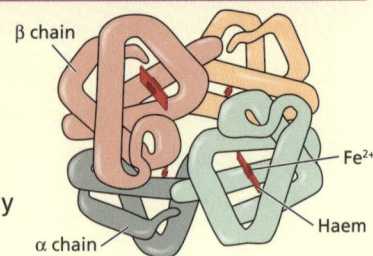

β chain, α chain, Fe²⁺, Haem

Oxyhaemoglobin dissociation curves

Haemoglobin can combine with oxygen when the concentration of oxygen is high. This forms **oxyhaemoglobin**. However, the reaction is reversible and, in low concentrations of oxygen, the oxygen is released. This binding and release can be plotted on an **oxyhaemoglobin dissociation curve**.

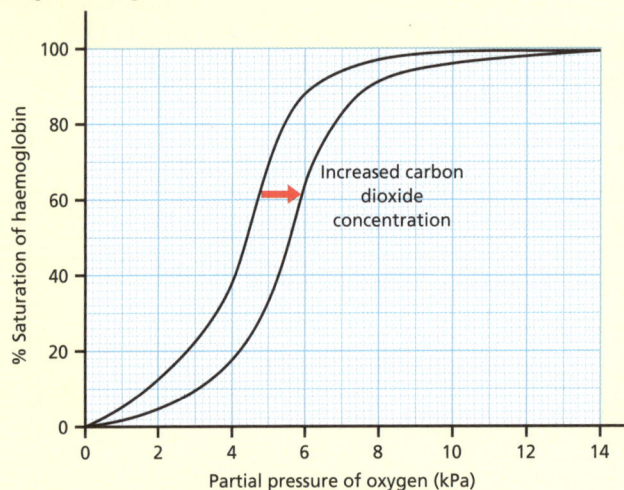

Graph: % Saturation of haemoglobin (y-axis, 0 to 100) vs Partial pressure of oxygen (kPa) (x-axis, 0 to 14). Arrow labelled "Increased carbon dioxide concentration".

The dissociation curve shows several important features:

- The shape is sigmoid, which means the middle portion is steep; a small change in concentration (partial pressure) of oxygen causes a large loading or unloading of oxygen.
- This curve shape is a result of the haemoglobin molecule changing shape when each molecule of oxygen binds so that it is easier for the next molecule to bind.
- When the concentration of carbon dioxide in solution increases, this lowers the affinity of haemoglobin for oxygen and so the curve shifts to the right. This decrease in affinity is called the **Bohr shift** and is important for the release of oxygen in respiring tissues.

✔ **Summary**

Haemoglobin and oxygen carriage

(1) How many atoms of oxygen can combine with one molecule of haemoglobin? Tick **one** box.　　　　[1]

2 ☐　　　4 ☐　　　6 ☐　　　8 ☐

(2) Tick **one** box to show which of these statements about haemoglobin are correct.　　　　[1]

1. Haemoglobin has a tertiary structure but no quaternary structure.

2. The polypeptide chains in haemoglobin combine reversibly with oxygen.

3. There are four haem groups in each haemoglobin molecule.

1, 2, 3 and 4 ☐　　　2 and 3 only ☐　　　2 only ☐　　　3 only ☐

(3) Look at the graph of the oxygen dissociation curve on page 54.

a) Calculate the change in percentage saturation of haemoglobin when blood moves from an area where the partial pressure of oxygen is 6 kPa to an area where it is 4 kPa.　　　　[2]

b) Calculate the change in percentage saturation of haemoglobin when blood moves into an area where the partial pressure of oxygen is 6 kPa to an area where it is also 6 kPa, but where the concentration of carbon dioxide is higher.　　　　[2]

c) Discuss the importance of the change in the saturation calculated in part b).　　　　[3]

(4) The graph shows oxygen dissociation curves for two different types of haemoglobin. Fetal Hb is found in babies' blood whilst they are in the uterus. Maternal Hb is in the mother's blood.

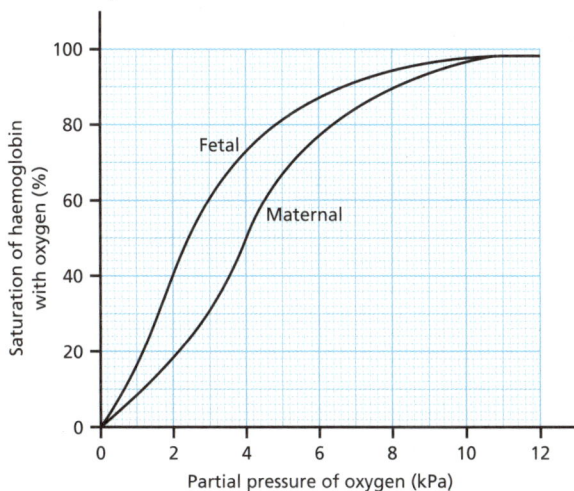

a) Maternal Hb that is 100% saturated with oxygen can carry 1.4 ml of oxygen per gram of haemoglobin.

Calculate the volume of oxygen in 1 gram of maternal Hb at an oxygen partial pressure of 4 kPa.　　　　[2]

b) Calculate the volume of oxygen carried by 1 gram of fetal Hb at an oxygen partial pressure of 4 kPa.

Assume that fetal Hb has the same total carrying capacity as maternal Hb.　　　　[2]

c) Complete these sentences by writing the correct words in the gaps.　　　　[3]

At most partial pressures, fetal haemoglobin is .... saturated with oxygen than maternal Hb.

This is because fetal Hb has a higher .... for oxygen than maternal haemoglobin.

Therefore, this allows more oxygen to pass to the fetus from the mother across the ....

Circulation of blood and the heart

Key words

Key questions

In which direction does blood travel in arteries?

Why is it important that blood in each side of the heart is kept separate?

What causes the valves in the heart to open?

Blood circulation

The blood in the human circulation passes through the heart twice in each circuit. It is pumped to the lungs to be oxygenated and then returns to the heart to be pumped to the rest of the body.

- The blood vessel passing blood to the lungs is the **pulmonary artery** and the blood returns in the **pulmonary vein**.
- Blood is pumped to the rest of the body in the **aorta** and is returned in the **venae cavae**.
- Blood from the aorta passes to the kidneys in the **renal arteries** and returns to the vena cava in the **renal vein**.
- **Coronary arteries** leave the aorta to take blood to the heart muscle.

Structure of the heart

The heart is really two pumps. The two sides are separated by a septum.

- The right side pumps deoxygenated blood to the lungs.
- The left side pumps oxygenated blood to the body.
- Valves prevent blood flowing backwards from the ventricles to the atria and from the aorta to the ventricles.
- Tendons prevent some of the valves from inverting.

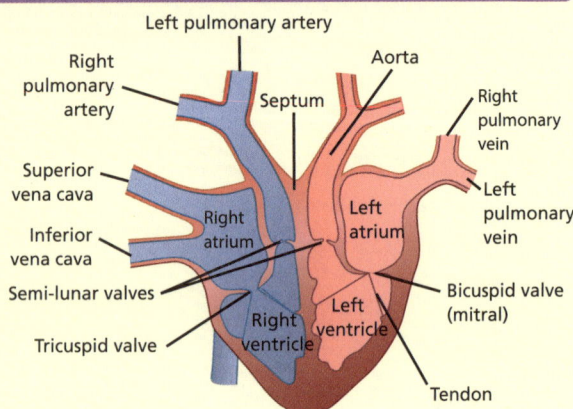

Labels: Left pulmonary artery, Right pulmonary artery, Septum, Aorta, Right pulmonary vein, Superior vena cava, Left atrium, Left pulmonary vein, Inferior vena cava, Right atrium, Bicuspid valve (mitral), Semi-lunar valves, Left ventricle, Tricuspid valve, Right ventricle, Tendon

Cardiac cycle

The four chambers of the heart pump blood by a series of contractions (**systole**) and relaxation (**diastole**):

- When the atria are full of blood, they contract, increasing the pressure of the blood and forcing open the valves into the ventricles. Blood flows into the relaxed ventricles.
- The atria then enter diastole and the ventricles enter systole.
- This increases the pressure of blood in the ventricles, shutting the valves to the atria and forcing blood into the arteries.
- The ventricles then relax and the higher pressure in the arteries shuts the semilunar valves.

This sequence of events lasts about 0.7 seconds and is called the **cardiac cycle**.

The graph shows these pressure changes for the left side of the heart.

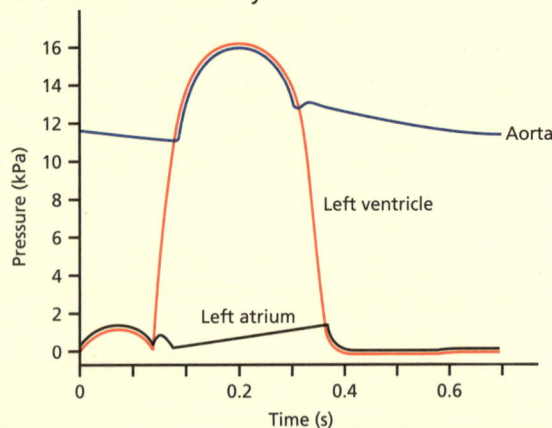

Graph: Pressure (kPa) against Time (s), showing Aorta, Left ventricle, and Left atrium curves.

Summary

RETRIEVE

Circulation of blood and the heart

1 Which two blood vessels contains deoxygenated blood? Tick **one** box. [1]

1. Vena cava **2.** Aorta **3.** Pulmonary artery **4.** Pulmonary vein

1 and 4 ☐ 1 and 3 ☐ 3 and 4 ☐ 2 and 3 ☐

2 The diagram shows the human heart viewed from the front.

Identify the structures labelled A, B, C and D. [4]

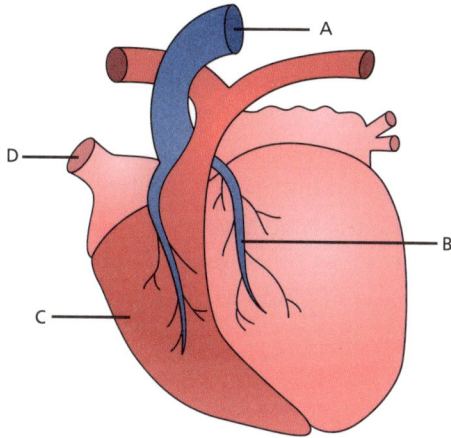

A = ...

B = ...

C = ...

D = ...

3 The graph shows the pressure changes during the cardiac cycle for the right side of the heart.

a) Compare the pressure values shown for the right ventricle with those shown for the left ventricle in the graph on page 56. Explain the difference in the values. [3]

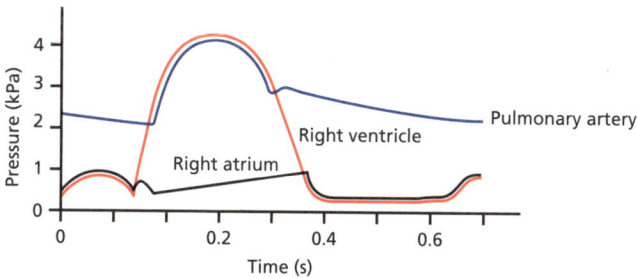

b) On the graph showing the pressures for the right side, label the position of these events:

A = opening of semilunar valve **B** = closing of semilunar valve

C = closing of tricuspid valve **D** = opening of tricuspid valve [4]

4 The diagram shows the events in one cardiac cycle of an individual.

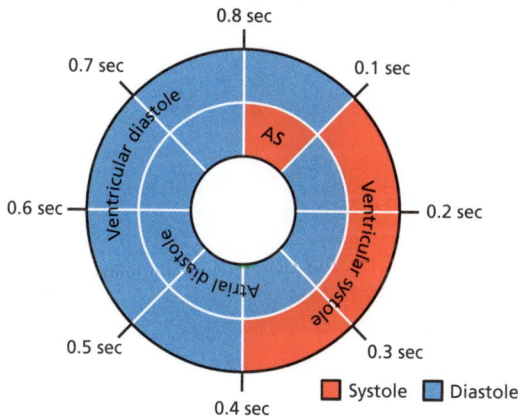

a) Calculate the heart rate for this person in beats per minute. [2]

b) Calculate the percentage of the cycle that is spent in atrial systole. [2]

c) Using the diagram, state the approximate time when the semilunar valves close. Explain your answer. [3]

Blood vessels and tissue fluid

Key words

Key questions

Why do veins have valves?

What is the function of smooth muscle in arterioles?

What causes tissue fluid to be reabsorbed?

Blood vessels

The human circulation is a closed system so relies on blood vessels to transport the blood. There are different types of blood vessels, adapted for their function.

Arteries transport blood away from the heart under pressure. They have:
- thick walls with collagen fibres to prevent bursting (blood is under high pressure)
- large amounts of elastic tissue in the tunica media so that the arteries can recoil to keep the blood moving when the heart relaxes
- smooth muscle fibres, especially in the smaller arteries, to regulate blood flow.

Capillaries carry the blood through the tissues. They have:
- thin walls made of a single layer of endothelial cells that allows exchange with the tissues
- very narrow lumen so that blood flow is slow, allowing time for exchange.

Veins carry blood back to the heart. They have:
- thin walls with less collagen as the blood pressure is much lower
- valves to prevent the backflow of blood as the pressure is low.

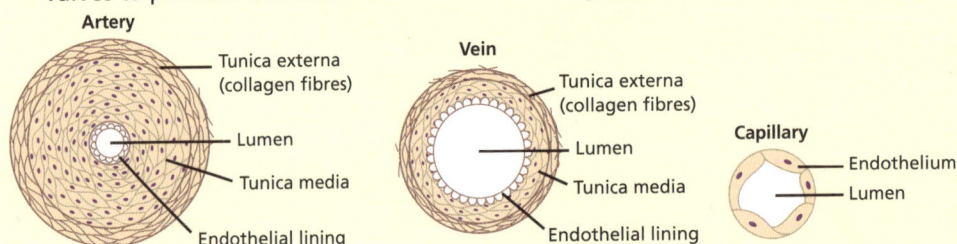

Arteries and veins are connected via small vessels called **arterioles** and **venules**. Arteries branch into arterioles, and the smooth muscle in the arterioles can contract or relax to adjust blood flow. Venules carry blood from the capillaries to the veins.

Tissue fluid

All tissues contain a network of capillaries that supply the cells with nutrients and remove wastes. This is done by the formation of **tissue fluid**:

Capillary bed

- Hydrostatic pressure in the capillaries forces fluid out of blood vessels.
- This liquid is like plasma but has a lower protein content and is called tissue fluid.
- The fluid bathes the cells, nutrients enter the cells, and waste products enter the tissue fluid.
- At the venous end of the capillary bed, the hydrostatic pressure of the blood drops.
- The water potential in the blood is low, due to the proteins in the blood, so the tissue fluid is drawn back into the capillaries by osmosis.

The tissues also have blind ending tubes (closed at one end) called **lymph vessels**. Some of the tissue fluid is absorbed into the lymph vessels and is transported through the lymphatic system. It returns to the blood near the heart, where a large lymph vessel called the thoracic duct joins with a vein.

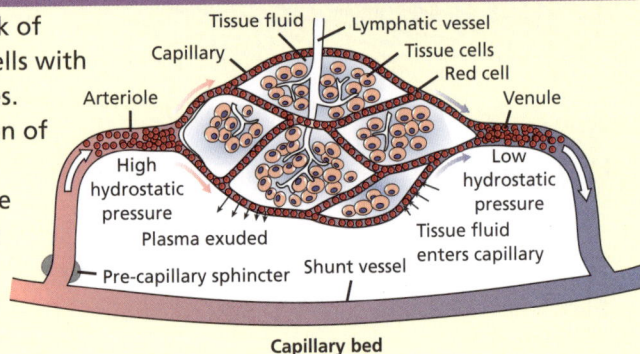

Summary

Blood vessels and tissue fluid

1) The graph shows how the pressure changes as blood passes through the five different types of blood vessel.

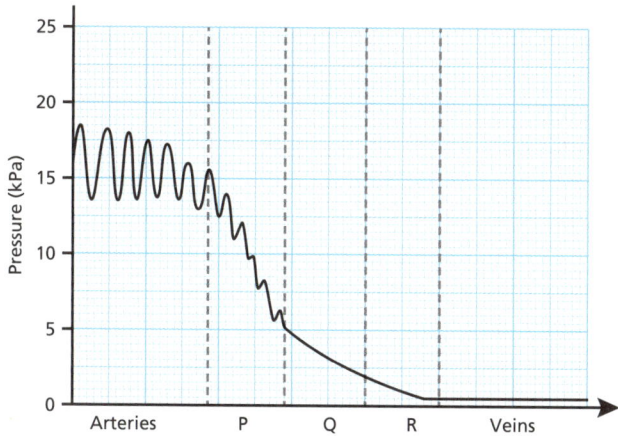

a) Identify the type of blood vessels P, Q and R. [3]

P = _____

Q = _____

R = _____

b) Give the maximum range of pressures in the arteries. [1]

....

c) Compare the pressure changes in blood vessels P with the pressure changes in blood vessels Q. [2]

....

....

d) Explain why the pressure in the arteries does not drop to zero when the heart ventricles relax. [2]

....

....

2) The diagram shows a capillary.

Diameter 8 µm

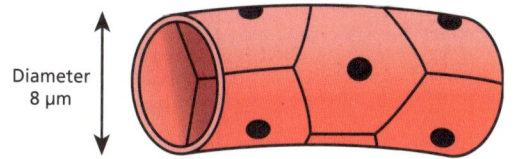

Explain how the capillary is adapted for its function. [3]

....

....

....

....

3) The diagram shows the formation of tissue fluid in a capillary bed.

Capillary containing blood

Blood hydrostatic pressure = 4.3
water potential = −3.3

Tissue fluid hydrostatic pressure = 1.1
water potential = −1.3

Cells of tissue

X

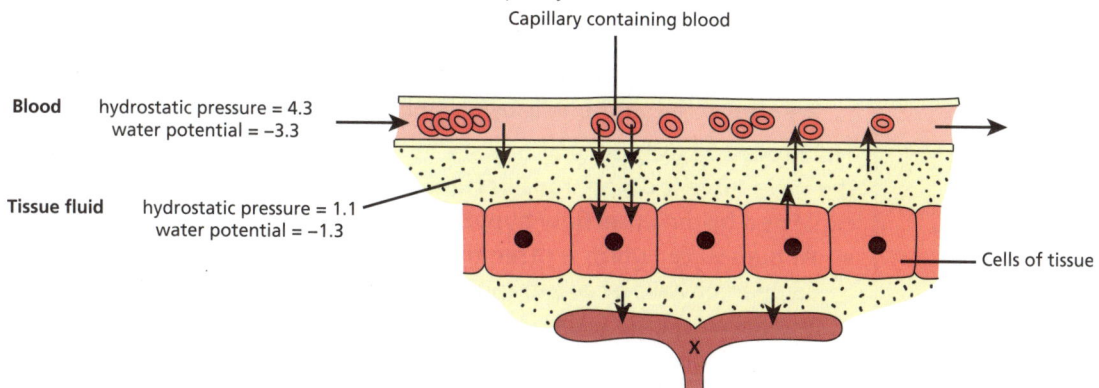

a) Use the data in the diagram to explain why tissue fluid is formed at the arterial end of the capillary bed. [3]

....

....

b) Name **two** substances that pass from the tissue fluid to the cells. [2]

....

c) Identify liquid X and describe how it is returned to the blood. [3]

....

....

Transport in plants

Key words

Key questions

Why do xylem vessels contain lignin?

What is cohesion?

Why do sieve tube cells need companion cells?

Xylem

Plants have two specialised tissues for the transport of substances: xylem and phloem. Xylem transports water and minerals from the roots to the leaves.

Xylem contains xylem vessels. These are dead cells that are adapted for the transport of water:

- They have cells end-to-end that have end walls broken down to form continuous tubes.
- The cells have lost their cytoplasm so there is less resistance to flow.
- The cell walls are strengthened with **lignin** to prevent the walls collapsing inwards.
- There are pits in the walls to allow the sideways movement of water.

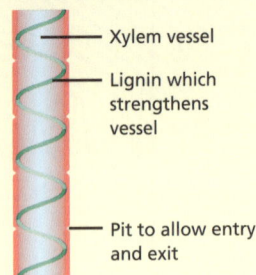

Xylem vessel
Lignin which strengthens vessel
Pit to allow entry and exit

Water is drawn into the root hair cells from the soil by osmosis. It then passes across the cortex until it enters the xylem vessels in the centre of the root. Water is then drawn up xylem according to the **cohesion-tension theory**:

- Water evaporates from mesophyll cells in the leaves and diffuses out of stomata.
- This process is called **transpiration** and it sets up a water potential gradient, drawing water out of the xylem vessels in the leaves.
- Hydrogen bonds between the water molecules in the xylem cause them to be attracted to each other (**cohesion**).
- The water molecules are therefore pulled up the xylem vessels from the roots.
- There is also a force called **root pressure**, which pushes water up from the roots due to the active transport of minerals into the xylem.

Phloem

Phloem tissue is responsible for the movement of dissolved food materials, such as sucrose, through the plant. The two main types of living cells in this tissue are **sieve tube cells** and **companion cells**.

- Sieve tube cells are arranged end-to-end, with the end walls forming sieve plates with pores in them.
- The sieve plates allow movement of solutes but provide some resistance to the cells bursting.
- There is limited cytoplasm in the companion cells, so there is less resistance to flow.
- Companion cells next to the sieve tube cells carry out metabolic processes for the sieve tube cells and are connected to the cells by **plasmodesmata**.

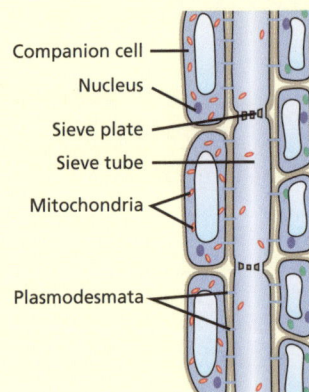

Companion cell
Nucleus
Sieve plate
Sieve tube
Mitochondria
Plasmodesmata

The process of transporting organic molecules through the sieve tubes is called **translocation** and can be explained by the **mass flow hypothesis**:

- At the source, which is often the leaves, sucrose molecules are loaded into the companion cells by a co-transporter and they then diffuse into the sieve tubes.
- This decreases the water potential in the sieve tubes, drawing in water by osmosis.
- The build-up of hydrostatic pressure forces the contents of the sieve tubes along to the sink, which is often actively growing areas or storage organs.
- At the sink, sucrose and other organic molecules are moved out by active processes.

Translocation has been studied by using radioactive isotopes to follow movement of molecules and ringing experiments, involving removing circles of phloem around the stem.

Summary

Transport in plants

1. Complete the table by adding ticks (✓) and crosses (✗) to show the properties of xylem vessels and phloem sieve tubes. [5]

	Xylem vessels	Sieve tube cells
Living cells		
Contains lignin		
Transports substances both up and down the stem		
Has end walls		
Contains some cytoplasm		

2. The diagram shows an experiment using a potted plant. Waterproof paper was inserted between the xylem and phloem in an area of the stem. The plant was watered with a solution containing radioactive K⁺ ions and the distribution of the ions recorded in different parts of the stem. The results are shown in the table.

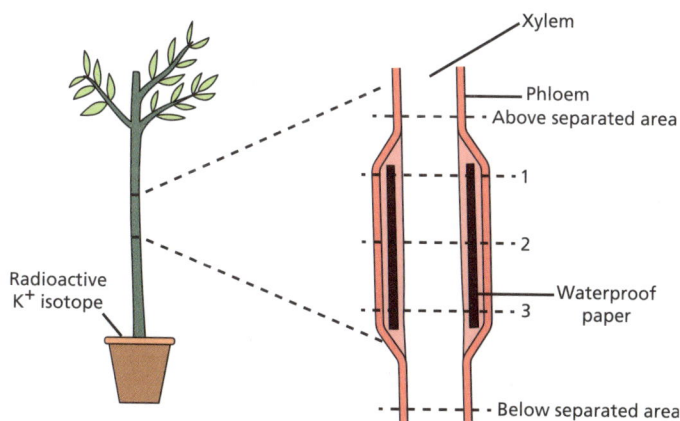

Area of the stem	Radioactivity (in arbitrary units)	
	phloem	xylem
Above separated area	42	45
1	10	78
2	0.5	88
3	15	72
Below separated area	43	46

a) The background radiation in the room was 0.5

Explain how the results show that mineral ions are transported in the xylem. [2]

b) Explain how the results show that minerals can usually move between the xylem and phloem. [2]

c) State the route that ions can take to move out of the xylem so that they can enter the phloem. [1]

3. The diagram shows a model used to explain how substances move along the phloem.

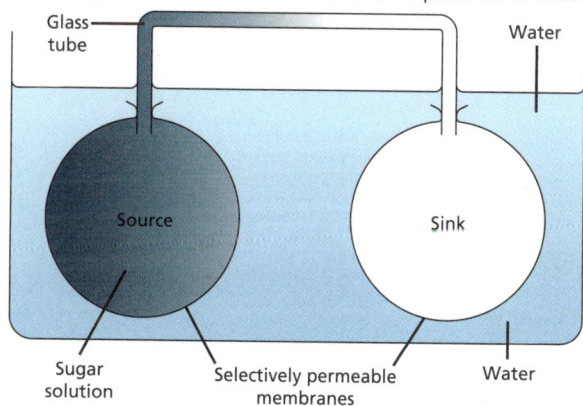

a) What does the glass tube in the model represent? [1]

b) Give the name of one part of a plant that often acts as a source and give the name of one part that is often a sink. [2]

source =

sink =

c) Explain how this model demonstrates the movement of substances from the source to the sink. [3]

DNA, genes and chromosomes

Key words

Key questions

Why is DNA coiled and supercoiled?

What is a locus?

What is the difference between introns and exons?

DNA and chromosomes

DNA is the genetic material. In eukaryotes most of it is in the nucleus in chromosomes. DNA molecules are very long, linear and associated with proteins, called **histones**. Together, a DNA molecule and its associated proteins form a chromosome. The long DNA molecules are coiled and supercoiled to allow them to fit in the nucleus.

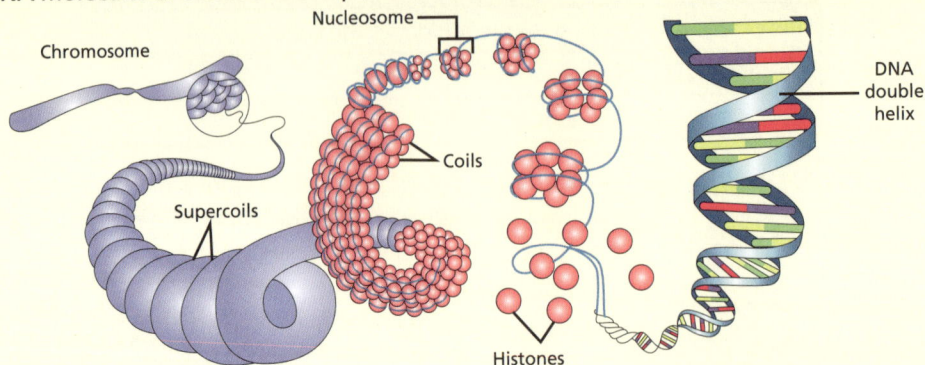

In prokaryotic cells, DNA molecules are short, circular and not associated with proteins.

Mitochondria and chloroplasts in eukaryotic cells also contain DNA which, like the DNA of prokaryotes, is short, circular and not associated with proteins. This DNA codes for a small number of proteins.

Genes and the genetic code

A gene is a base sequence of DNA that codes for either:

- the amino acid sequence of a polypeptide, or
- an RNA molecule (including ribosomal RNA and tRNAs).

Each gene occupies a fixed position, called a **locus**, on a particular chromosome.

The order of DNA bases forms the **genetic code**. The genetic code is:

- a triplet code, with each three bases coding for one amino acid
- universal, as the same triplet codes for the same amino acid in almost all organisms
- non-overlapping, as each triplet of bases is discrete
- degenerate, as some amino acids are coded for by more than one triplet of bases.

In eukaryotes, much of the nuclear DNA does not code for polypeptides. This DNA can be in different places:

- Between genes there are non-coding multiple repeats of base sequences.
- Within a gene there are one or more non-coding sequences, called **introns**, and these separate **exons** that code for amino acid sequences.
- There are stop and start codons at the beginning and end of genes that are bases marking each end of the gene.

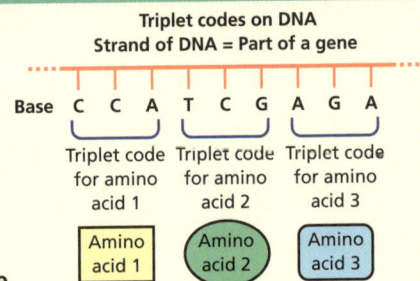

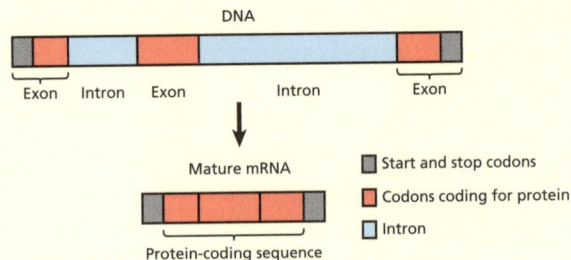

Summary

DNA, genes and chromosomes

1. Which of these statements correctly describes the genetic code? Tick **one** box. [1]

 1. The final base of each triplet acts as the first base for the next triplet.
 2. Each amino acid is only coded for by one specific base triplet.
 3. Some base triplets do not code for amino acids and so mark the end of genes.

 1, 2 and 3 ☐   2 and 3 only ☐   1 only ☐   3 only ☐

2. Complete the table by adding ticks (✓) and crosses (✗) to show the properties of DNA from different sources. [4]

	Eukaryotic nuclear DNA	Mitochondrial DNA	Bacterial DNA
Associated with histone proteins			
Codes for all the proteins made in the cell			
Contains introns			
Enclosed by a double membrane			

3. DNA contains four different types of bases. There are 20 amino acids found in proteins.

 a) Show by calculations that a triplet code is the minimum number of bases that is needed to code for all 20 amino acids. [2]

 b) Explain why a triplet code allows the genetic code to be degenerate. [2]

4. Read this passage about insulin.

 Insulin is a protein hormone made of two polypeptide chains. One of the chains has 21 amino acids and the other has 30. The gene that codes for insulin is found on chromosome 11. This INS gene has three exons and two introns. Insulin is produced as proinsulin and this is then cut by enzymes to make two chains. The two chains are joined by disulphide bonds to form the insulin molecule.

 For many years, pig insulin was used to treat diabetes. Cow insulin differs from human insulin by three amino acids and pig insulin differs by only one amino acid. Genetically engineered human insulin is now used, which is produced by bacteria that have had the human gene inserted.

 Use the passage to answer the following questions.

 a) Calculate how many nucleotide pairs are actually involved in coding for the amino acids that make up the insulin molecule. [2]

 b) The INS gene is found on chromosome 11.
 Name the term used to describe the position of a gene. [1]

 c) Name the parts of the INS gene that code for the insulin protein. [1]

 d) Explain why pig insulin is more suitable than cow insulin in treating diabetes in humans. [2]

 e) Bacteria that have the human INS gene inserted produce human insulin.
 Explain which feature of the genetic code makes this possible. [2]

Protein synthesis

Key words

Key questions

Why does DNA have to unwind and unzip in translation?

Why is mRNA needed in protein synthesis?

What is an anticodon?

Transcription

Each cell contains the complete set of genes for that organism. The set of genes is called the **genome**.

All the proteins that a cell can produce at a certain time is called the **proteome**.

The first step in protein synthesis is **transcription**. This involves the production of an mRNA molecule using the DNA for the specific gene as a template:

- The portion of the DNA containing the gene unwinds and the hydrogen bonds between the DNA bases break, causing the two strands of the double helix to separate.
- RNA nucleotides pair up with DNA nucleotides on the template strand by complementary base pairing.
- Uracil pairs with adenine and cytosine pairs with guanine.
- RNA polymerase catalyses the formation of phosphodiester bonds between adjacent RNA nucleotides, forming a complementary mRNA molecule.
- The process ends when RNA polymerase reaches a stop base triplet, detaches from DNA and terminates transcription.
- mRNA detaches from the DNA and the DNA rewinds into its double helix structure.

In prokaryotes, transcription results directly in the production of mRNA from DNA. However, in eukaryotes, transcription results in the production of pre-mRNA. This contains sections of introns that have been transcribed. This pre-mRNA is then spliced to remove the introns and form mRNA.

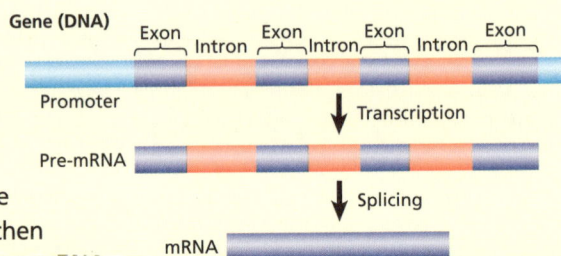

Translation

Translation is the production of polypeptides using the information carried by mRNA. The DNA cannot leave the nucleus, so mRNA carries the information in the form of base triplets called **codons**.

The process of translation occurs on the ribosomes and involves base triplets or **anticodons** on tRNA molecules base pairing with mRNA codons. Each tRNA molecule has an amino acid attached that is specific to the anticodon. The correct amino acid is brought into place by the tRNA molecule.

Peptide bonds are formed between the amino acids using enzymes on the ribosome and energy from ATP.

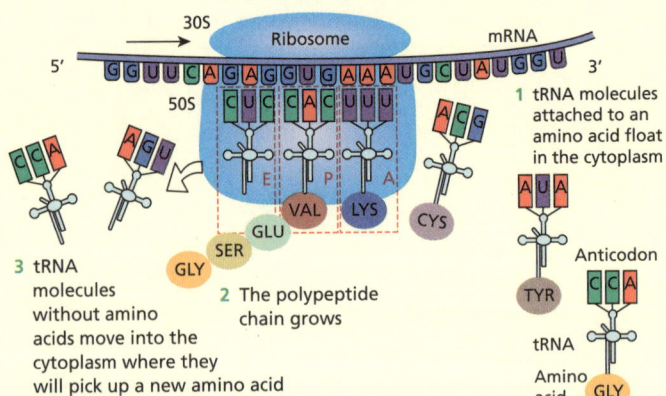

Summary

Protein synthesis

① Which of these statements about protein synthesis are correct? Tick **one** box. [1]

1. Splicing of mRNA is not required in bacteria.

2. The end products of translation are polypeptides and ATP.

3. DNA polymerase and RNA polymerase are both required for protein synthesis.

1, 2 and 3 ☐   2 and 3 only ☐   1 and 3 only ☐   1 only ☐

② Complete the table by adding ticks (✔) and crosses (✗) to show which statements apply to transcription and translation. [4]

	Transcription	Translation
Involves base pairing		
Involves enzymes		
Involves DNA		
Breaks peptide bonds		

③ The list below gives five processes involved in protein synthesis in eukaryotes.

1. Production of pre-mRNA

2. Unwinding of DNA

3. Base pairing between codons and anticodons

4. Base pairing between RNA nucleotides and DNA

5. Splicing of mRNA

Write the numbers 1–5 in the boxes to show the correct order of these processes. [4]

☐ ☐ ☐ ☐ ☐

④ The diagram shows the mRNA codons that code for each amino acid. The amino acids are shown on the outer ring, represented by three letters. The inner circle gives the first base of each triplet, the next ring gives the second base of the triplet and the third ring gives the last base. For example, GCA codes for Ala.

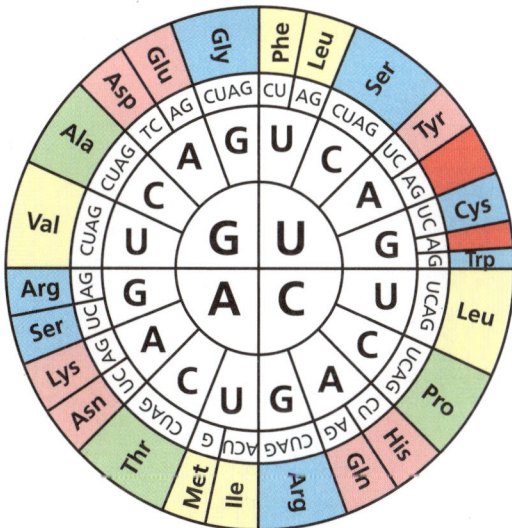

a) How many different codons code for Asp? [1]

b) Which anticodon would be found on the tRNA molecule that carries the amino acid Met? [1]

c) A length of mRNA has this base sequence: AUGAGGUUUCAC

Give the amino acid sequence that this would produce after translation. [1]

d) The triplet codes UGA, UAG and UAA are called stop or termination codons, marking the end of the gene.

Use the diagram to explain why they act in this way. [2]

Meiosis

🔓 Key words

❓ Key questions

What does diploid mean?

At which stage of meiosis do homologous chromosomes pair up?

Why does the centromere divide in meiosis II?

The function of meiosis

Meiosis is a type of cell division that reduces the chromosome number by half.

In the cells of eukaryotic organisms, there are two copies of each chromosome. This is called the **diploid** state (2n) and the pairs are called **homologous chromosomes**.

Meiosis is used to produce gametes that contain one copy of each homologous chromosome. This is called the **haploid** state (n). The gametes can then fuse to produce a diploid zygote. Unlike mitosis, there are two divisions in meiosis and so four haploid cells are produced, rather than two diploid cells.

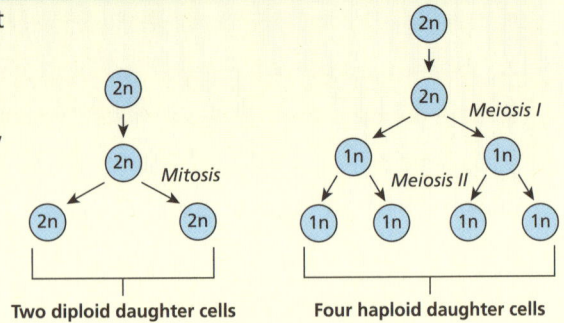

Two diploid daughter cells

Four haploid daughter cells

The process of meiosis

As in mitosis (page 36), there is an interphase before meiosis, during which the chromosomes are replicated. The stages of meiosis have the same names as in mitosis but have I or II after the name to indicate whether they are in the first or second meiotic division.

The stages of meiosis are shown in the diagram below:
- In meiosis I, the homologous chromosomes pair up and one chromosome from each pair moves to opposite poles of the cell.
- In meiosis II, the centromeres divide and the chromatids from each homologous chromosome move to opposite poles.
- Cytokinesis occurs in the same way as in mitosis, with the cell membrane pinching in if it is an animal cell or the formation of a new cell wall if it is a plant cell.

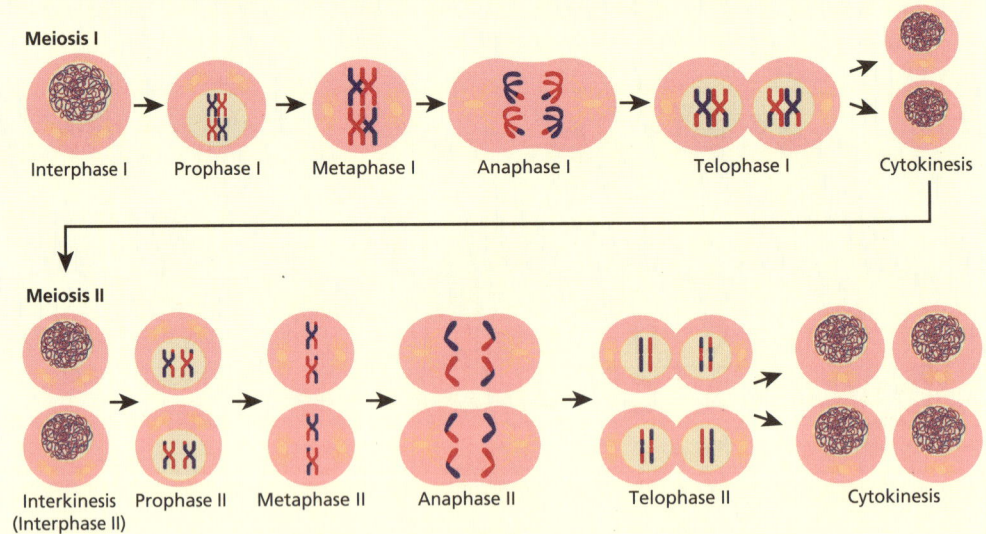

Meiosis I

Interphase I   Prophase I   Metaphase I   Anaphase I   Telophase I   Cytokinesis

Meiosis II

Interkinesis   Prophase II   Metaphase II   Anaphase II   Telophase II   Cytokinesis
(Interphase II)

In plants, meiosis occurs in the production of the female gametes in the ovary of a flower and in the pollen grains that contain the male gamete.

Some single-celled organisms are haploid and they produce a diploid zygote. This then divides by meiosis to produce haploid individuals.

✔ Summary

Meiosis

① The list below gives four processes involved in meiosis.

   1. Chromatids move apart.

   2. Homologous chromosomes pair up.

   3. Nuclear membrane first disappears.

   4. Centromeres divide.

Write the numbers 1–4 in the boxes to show the correct order of these processes. [3]

   ☐ ☐ ☐ ☐

② Complete the table by adding ticks (✓) and crosses (✗) to show which statements apply to meiosis and to mitosis. [4]

	Mitosis	Meiosis
Homologous chromosomes pair up		
Spindle fibres shorten		
Produces two cells		
Reduces the chromosome number		

③ The diagram shows the contents of a nucleus of a cell in interphase I, before the DNA has replicated.

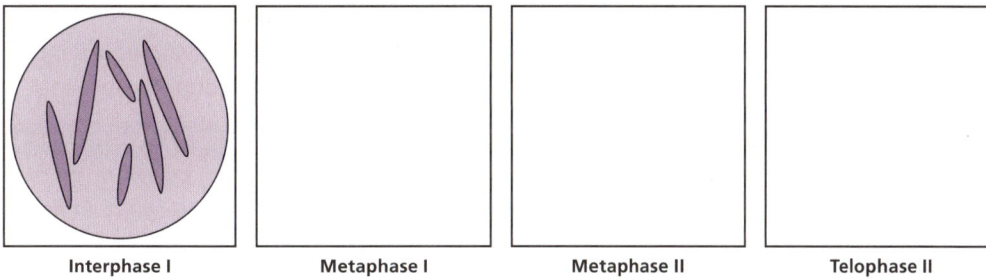

Interphase I   **Metaphase I**   **Metaphase II**   **Telophase II**

  a) Give the diploid number for this cell. [1]

  b) In each box, draw the appearance of the nucleus for one cell for the three labelled stages of meiosis. [6]

  c) Describe what will happen after the stage telophase II. [2]

  d) Describe how the contents would appear different if the cell was viewed in metaphase of mitosis, compared to metaphase I of meiosis. [2]

④ The graph shows the DNA content of a cell before, during and after meiosis.

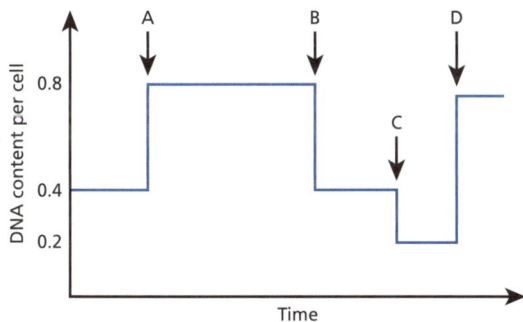

  a) Name the processes occurring at these points. [3]

     A = ...

     B and C = ...

     D = ...

  b) Mark on the graph with an X where meiosis II is occurring. [1]

Mutation and variation

Key words

Mutation

Mutations are permanent changes in the genetic material of a cell. There are two types of mutations.

Gene mutations are changes in the base sequence in one gene. This occurs when the DNA is replicated before mitosis or meiosis. A gene mutation may involve a change to a single base (point mutation) or to a section of the gene. The types of gene mutations are:

- a **substitution** mutation that involves one base being changed for another
- an **addition** mutation that involves an extra base or bases being inserted
- a **deletion** mutation that involves one or more bases being removed from the gene.

Substitution mutations may result in a different amino acid being coded for. This may affect the shape and functioning of the protein produced by the gene. However, due to the degenerate nature of the genetic code, the base triplet could still code for the same amino acid. Also, if a similar amino acid is coded for, this may not affect the structure of the protein. These mutations are **neutral mutations**. Addition or deletion mutations will usually change all the amino acids that are coded for further along the gene. This is called a **frameshift mutation**.

Chromosome mutations are changes in sections of a chromosome or whole chromosomes. This may occur during cell division if the chromosomes or chromatids are not separated correctly at anaphase. This is called **non-disjunction** and can produce a cell with an extra chromosome and another with a chromosome missing.

Mutations occur spontaneously at a low rate but the rate can be increased by mutagenic agents such as UV light and chemicals in tar from tobacco smoking.

Key questions

What is a gene mutation?

Why are frameshift mutations usually harmful?

What is a chiasma?

Variation

Mutations produced by errors in DNA replication can produce new alleles of genes. However, meiosis and fertilisation can recombine these alleles in different combinations. This produces variation and can occur in different ways.

Crossing over occurs during prophase I of meiosis. This happens when homologous chromosomes pair up and become attached at places called chiasmata (singular: chiasma). When the chromosomes move apart, a chromatid may break and rejoin with the other chromatid. This causes alleles to be swapped between the homologous chromosomes.

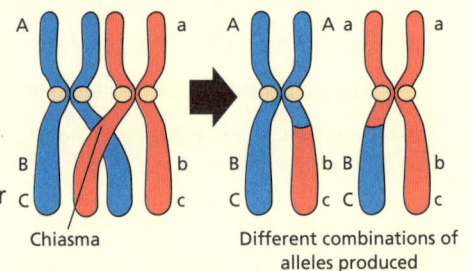

Chiasma

Different combinations of alleles produced

Independent segregation occurs during metaphase I of meiosis. Homologous chromosomes can pair up in different orientations at metaphase I. This results in different combinations in the gametes. The diagram shows the situation with two pairs of homologous chromosomes. With 23 pairs, there would be 2^{23} possible combinations.

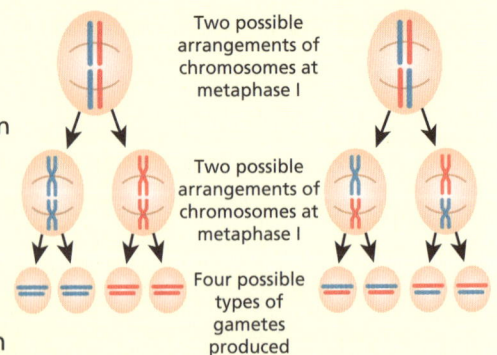

Random fusion of gametes at fertilisation also increases the possible variation in the offspring produced.

Two possible arrangements of chromosomes at metaphase I

Two possible arrangements of chromosomes at metaphase I

Four possible types of gametes produced

Summary

Mutation and variation

(1) Houseflies have six pairs of homologous chromosomes.

How many different combinations of chromosomes can be produced by meiosis as a result of independent segregation? Tick **one** box. [1]

6 ☐   12 ☐   36 ☐   64 ☐

(2) Colchicine is a chemical that disrupts the formation of spindle fibres in meiosis.

Which of these statements could be the effect of treating a cell with colchicine? Tick **one** box. [1]

1. Gene mutation

2. Chromosome mutation

3. Non-disjunction

4. Same sense mutation

1, 2, 3 and 4 ☐   2 and 3 only ☐   1 and 3 only ☐   2 and 4 only ☐

(3) The diagram shows two homologous chromosomes at the start of prophase 1 of meiosis. The alleles for two genes are shown on the chromosomes.

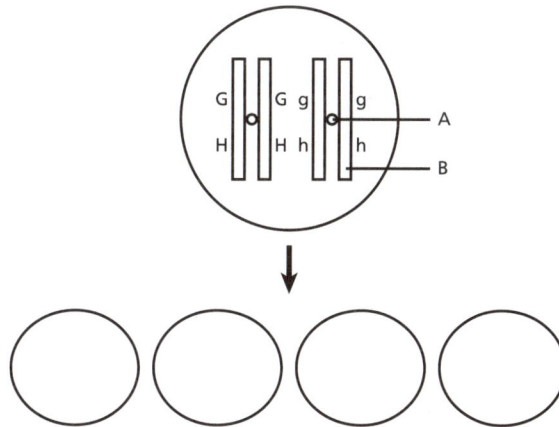

a) Identify the structures labelled A and B. [2]

A = ...

B = ...

b) In the four circles, draw the four possible gametes that could be formed at the end of meiosis. [4]

(4) This is the DNA base sequence for part of a gene: T A C A C C G A G C G A

To help answer these questions, use the diagram in question 4 on page 65.

a) Give the mRNA that would be produced by transcribing this sequence. .... [2]

b) Give the amino acid sequence that would be produced by translation. [1]

....

c) A mutation produces this new base sequence for this part of the gene: T A C A C A G A G C G A

i) Give the name of this type of gene mutation. .... [1]

ii) Identify the effect that this mutation would have on the amino acid sequence. [1]

....

iii) Explain why a mutation that changed the base to T instead of A would have a more serious effect on the protein produced. [2]

....

....

Natural selection

Genetic diversity and variation

The production of new alleles by mutation and their rearrangement into different combinations by meiosis produces variation between organisms. The environment can also cause variation in characteristics.

Continuous variation is variation where all intermediate values are possible, e.g. height. **Discontinuous variation** is where the individuals fit into discrete categories, e.g. blood groups.

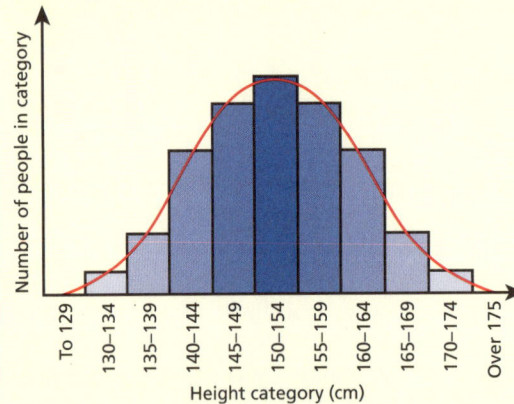

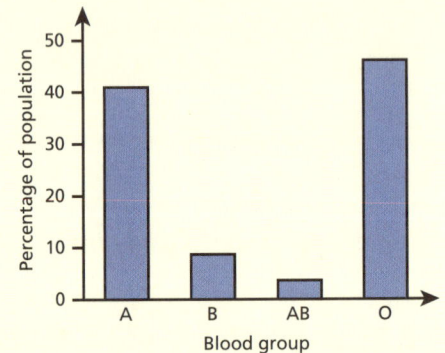

Key questions

What is the importance of mutation in producing genetic diversity?

Although continuous variation is more affected by the environment, both types of variation can be affected by the alleles of an individual.

The number of different alleles that are present in a population is called the **genetic diversity**. Genetic diversity in a population allows evolution to occur.

Natural selection

The idea that the characteristics of populations of organisms change over time is called **evolution**. For many years, scientists believed that evolution occurred but the mechanism was unknown. Charles Darwin published a possible mechanism in 1859 called **natural selection**. This has since been modified owing to modern genetic discoveries, but it is still based on Darwin's original idea:

What is meant by 'survival of the fittest'?

- Random mutation in individuals in a population can result in new alleles of a gene.
- Many mutations are harmful but the new allele of a gene might make the individual better suited to the environment; that makes it more likely to survive and reproduce.
- The advantageous allele is therefore passed on to members of the next generation.
- Over many generations, the new allele increases in frequency in the population.

Natural selection therefore makes populations better adapted to their environment. These adaptations may be:

- anatomical, involving the structure of organisms
- physiological, involving bodily processes
- behavioural, changing the responses of organisms to stimuli.

What happens if the genetic diversity of a population is very low?

Natural selection relies on genetic diversity in a population. So, a lack of genetic diversity can result in a population being unable to evolve to become adapted to changes in its environment. There are examples of this happening, such as the Hawaiian goose. Hunting reduced the population in Hawaii to about 30 individuals. Numbers have now increased to nearly 4000 but there is a lack of genetic diversity in the population.

Summary

Natural selection

1. Some rodents that live in deserts have evolved to spend more time underground during the day.

   Put a tick (✔) next to this type of adaptation. [1]

   Anatomical ☐   Behavioural ☐   Physiological ☐

2. A student investigated the ability of other students to be able to roll their tongues. They found that some students could roll their tongue and some could not. This ability did not seem to give the students any biological advantage.

   Put a tick (✔) next to the statement that describes this type of variation. [1]

   Continuous variation which is controlled by genes ☐

   Continuous variation which is controlled by the environment ☐

   Discontinuous variation which is controlled by genes ☐

   Discontinuous variation which is controlled by the environment ☐

3. Which of these factors are needed for natural selection to occur? Tick **one** box. [1]

   1. Genetic variation

   2. A stable environment

   3. Selection pressure

   1, 2 and 3 ☐   1 and 3 only ☐   2 and 3 only ☐   1 and 2 only ☐

4. The five statements A–E describe events in the process of natural selection.

   A. An advantageous allele is passed on.

   B. An allele increases in frequency in the population.

   C. Random mutation occurs.

   D. Organisms reproduce.

   E. The best-adapted organisms survive.

   Write the letters A to E in the correct boxes to show the order of these events in natural selection. [4]

   ☐ ☐ ☐ ☐ ☐

5. Rabbits were first introduced into Australia in 1788 and became widespread. In 1950, the myxoma virus was introduced to try to control rabbit numbers. This virus killed rabbits and reduced the population numbers. However, the rabbits have now adapted and numbers are increasing.

   a) Suggest why rabbit numbers increased rapidly when they were first introduced into Australia. [1]

   b) Use ideas about natural selection to explain how the rabbits became adapted so that their numbers are now increasing. [5]

Types of natural selection

Key words

Key questions

What phenotype is selected for in directional selection?

Directional selection

Directional selection is a type of natural selection that occurs when one extreme phenotype is favoured and selected for, rather than phenotypes from the other extreme.

This means that the allele controlling the favoured phenotype will increase in frequency in the population. This will cause a curve showing the distribution of phenotypes to shift towards one end.

A good example of directional selection can be seen in the spread of antibiotic resistance in bacteria.

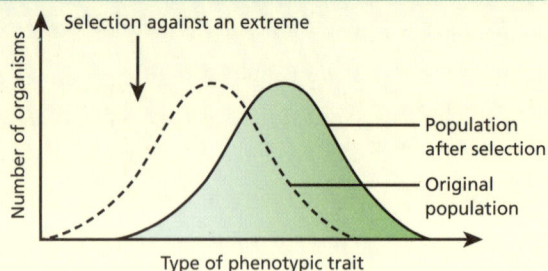

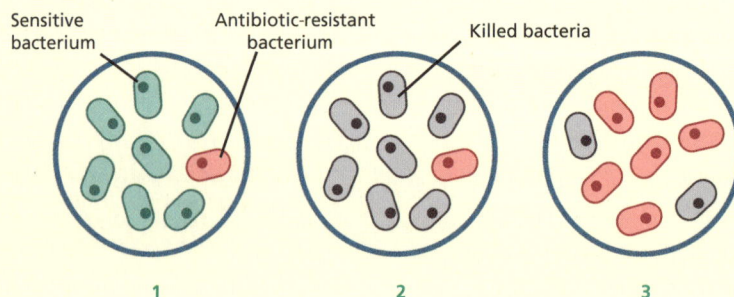

How is antibiotic resistance initially produced?

1. A random mutation can produce a bacterium that is resistant to antibiotics.
2. The use of antibiotics introduces a selection pressure, which kills any sensitive bacteria but allows the resistant bacteria to survive.
3. The resistant bacteria can reproduce and build up a strain of resistant bacteria.

Stabilising selection

Stabilising selection is a type of natural selection that occurs when an average or mean phenotype is favoured or selected. Phenotypes at the two extremes are selected against. This causes the distribution to become narrower.

This is probably the most common type of natural selection, occurring if the environment stays fairly constant.

What phenotype is selected for in directional selection?

Human birth weight is a good example of stabilising selection. Studying records of birth weights and death rates of babies show that those that have a medium birth weight are more likely to survive. Heavier babies have higher death rates as the birthing process becomes more challenging. Lighter babies are more likely to become ill.

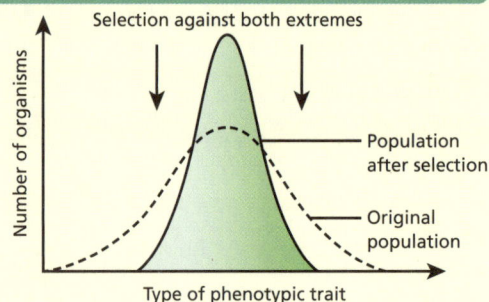

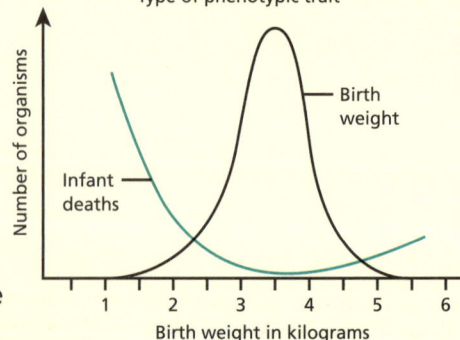

Summary

Types of natural selection

1. Which of these statements is true for stabilising selection? Tick **one** box. [1]

Only one narrow range of phenotype is selected for. ☐

Only phenotypes at one extreme of the distribution are selected for. ☐

Phenotypes in two specific ranges are selected for. ☐

The mean of the distribution of phenotypes is shifted. ☐

2. The paragraphs describe different features of plants and animals.

a) Cacti have sharp spines on their stems. The stems are fleshy and some animals like to eat them. This can be prevented by increasing the number of spines on the cactus. Cacti also have a parasitic insect that will lay its eggs in spines if the spines are close together.

Explain which type of selection controls the density of spines on cacti. [3]

....

....

b) A bird called the ground finch lives on one of the small Galápagos Islands. These birds eat seeds. In 1977 there was a drought on the island. Plants that are drought-resistant tend to have big seeds with thicker shells. Scientists noticed that there was a large increase in the size of the birds' beaks the year after the drought.

Explain which type of selection controls the size of the birds' beaks. [3]

....

....

c) Plants often compete with other plants for sunlight. Tall plants need to develop strengthening tissue to prevent them being damaged by wind.

Explain which type of selection controls the height of many plants. [3]

....

....

3. Snow geese are birds that lay eggs in nests on the ground. The female bird sits on the eggs to keep them warm and protects the young birds when the eggs hatch. Scientists studied a population of snow geese and recorded different aspects of their egg laying. The results are shown in the table.

Number of eggs in each nest	Number of nests with each number of eggs	Number of eggs that hatch	Percentage of eggs that hatch
2	3	2	33.3
4	36	108	
6	9	37	68.5
9	1	3	33.3

a) Calculate the missing value in the table. [2]

....

b) Which type of selection seems to act on the number of eggs laid by the snow goose? [1]

....

c) Suggest **two** reasons why the eggs in the nests with 9 eggs are selected against. [2]

....

d) The scientists only recorded the number of eggs that hatched.

Give a reason why it would be important to also study the percentage of young that survived to leave the nest. [1]

....

Classification

Key words

Key questions

Why is it important that organisms only mate within species?

What is the function of courtship?

Why is a phylogenetic classification system useful?

The species concept

The basic unit of classification is the **species**.

A species is defined as a group of organisms with similar characteristics that can interbreed and produce fertile offspring.

Members of some closely related species can mate. For example, horses and donkeys can breed and produce a mule.

Lions and tigers can mate, with tigons or ligers being the product. However, mules, tigons and ligers are sterile and are called **hybrids**. A tigon, which is produced by a male tiger and a female lion, is shown in the picture.

Members of closely related species can look quite similar and within a species there is often a lot of variation. It is therefore important that organisms can identify members of their own species to mate with. This is aided by **courtship behaviour**.

This can take various forms, including:

- sounds – birds, for example, sing in ways that are species-specific
- physical displays – examples include feathers in birds or rectal displays in monkeys
- dancing – common in birds such as the red-crowned cranes shown in the picture.

Classification

All life on Earth can be classified into groups. This process is called **taxonomy**. The groups are called **taxa** (singular: taxon).

Taxa are in a hierarchy in which smaller groups are placed within larger groups, with no overlap between groups. This is shown in the diagram.

The earliest method of classification was based on grouping by similar morphological (body plan) characteristics.

Modern technology now classifies organisms based on evolutionary origins and relationships. This is called a **phylogenetic** classification system. Analysis of protein structure, immunology and DNA sequencing are used in phylogenetics.

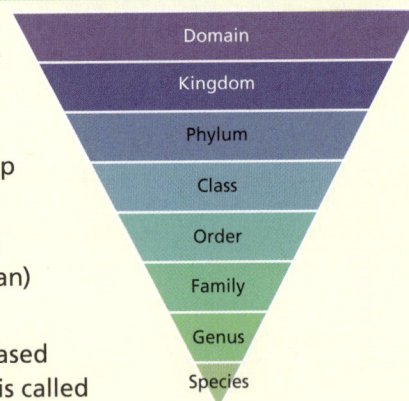

Domain
Kingdom
Phylum
Class
Order
Family
Genus
Species

Nomenclature

Nomenclature is the system of naming organisms. The system used now was devised by Carl Linnaeus in the 18th century and is called the **binomial system**.

Each organism has a scientific name that has two parts. The first is the name of the genus that contains the organism and the second part is the species name. For example, modern humans are called *Homo sapiens* and lions are called *Panthera leo*. By convention, the name is in italics when typed and the genus name starts with an uppercase letter.

Summary

Classification

1. Many species of bird share similar characteristics. In tropical rainforests, there are birds of paradise. There are many different species of these birds and they strongly defend their individual territories from each other. The forest is also particularly dense. Birds of paradise have very elaborate courtship behaviours, involving unique dances and songs.

   a) Explain why courtship behaviour is important to all birds. [3]

   b) Explain why courtship behaviour is particularly important to birds of paradise. [2]

   c) Give a reason why courtship songs are particularly important in tropical rainforests. [1]

2. Minke whales are marine mammals. There are two different species: one is called the northern minke whale.

   The table shows the classification of the northern minke whale, which has the scientific name *Balaenoptera acutorostrata*.

Domain	Eukaryota
Kingdom	Animalia
	Chordata
Class	Mammalia
Order	Artiodactyla
Family	Balaenopteridae
Species	

   a) Complete the table by filling in the gaps. [4]

   b) Explain why scientists usually use the scientific name for the northern minke whale rather than the common name. [2]

   c) Scientists have recently shown that northern minke whales can successfully mate with the other species of minke whale. Describe what would be produced from this mating. [2]

   d) Scientists have studied the DNA of common minke whales, dolphins, cows and pigs. The Venn diagram shows the number of shared and non-shared gene families in these species.

   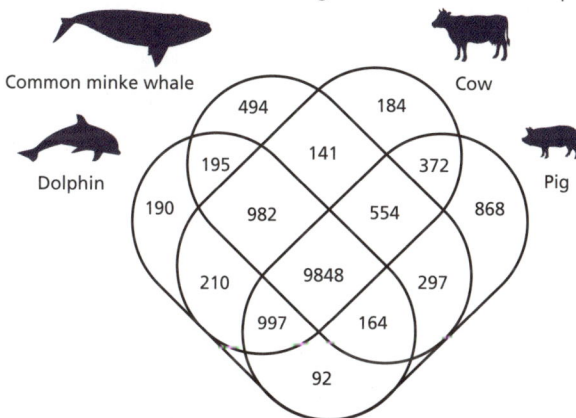

   i) Calculate the total number of gene families in the common minke whale. [1]

   ii) The table gives the percentage of the minke whale gene families shared by the minke whale and each of the other animals.

      Calculate the missing percentage in the table.

Animals	Percentage of gene families shared
whale and dolphin	88
whale and pig	
whale and cow	91

   [2]

   iii) Use the table to identify which animal the minke whale is most closely related to. [1]

   iv) What name is given to this type of evolutionary study? [1]

Species biodiversity

Key words

Key questions

What is the difference between species richness and species evenness?

What does an index of diversity value of 0.99 indicate?

Why is high biodiversity important in an area?

Species diversity and richness

Species richness is the number of different species in a community.

Species evenness is the number of individuals of each species (its abundance).

In a typical British woodland, there are many trees of different species. In an example, woodland A and woodland B both contain 100 trees. However, the numbers of each species are different:
Woodland A has 25 silver birch, 30 beech, 18 ash and 27 oak.
Woodland B has 75 silver birch, 5 beech, 2 ash and 18 oak.

The species richness in each area is the same, as woodland A and woodland B both have four different tree species. The species evenness, however, is different. Woodland A has a more even number of individuals from each species but woodland B is more dominated by silver birch.

Index of diversity

Species diversity gives an indication of the variety of organisms in a habitat. A measure of species diversity must consider both species richness and evenness.

A formula can be used to give an index of diversity d for species diversity of an area.

The formula is: $d = \frac{N(N-1)}{\sum n(n-1)}$

where N is the total number of organisms of all species and n is the total number of organisms of each species.

The box shows the calculations for the two woodlands in the above example. The larger d is, the higher the diversity in the area.

In the example, woodland A has a higher diversity as $d = 4.0$ compared with 1.7 in woodland B.

Using the example given for **woodland A:**
Silver birch = 25 × 24 = 600
Beech = 30 × 29 = 870
Ash = 18 × 17 = 306
Oak = 27 × 26 = 702
Total = 2478
$N(N-1) = 100 × 99 = 9900$ So d = **4.0**

Using the example given for **woodland B:**
Silver birch = 75 × 74 = 5550
Beech = 5 × 4 = 20
Ash = 2 × 1 = 2
Oak = 18 × 17 = 306
Total = 5878
$N(N-1) = 100 × 99 = 9900$ So d = **1.7**

Threats to diversity

A higher species diversity for a habitat means that the area is more likely to be stable and less likely to be affected by any environmental changes that might occur.

Farming practices can reduce biodiversity in a number of ways:
- Natural hedgerows between fields, which act as habitats for plants and animals, have been removed to convert many small fields into larger fields.
- A single crop is grown over a large area, forming a monoculture.
- Widespread use of insecticides (to kill insect pests) and herbicides (to kill weeds) has reduced biodiversity.
- Areas of naturally high biodiversity, such as forests, may be cleared to use as farmland.

Conservation schemes set up to try to reduce the impact of farming on biodiversity:
- Organic farming does not permit the use of insecticides or herbicides.
- Strips of land around fields have been left uncultivated to encourage wildlife.
- Areas of high biodiversity are protected by law.

Summary

Species biodiversity

1 Which of these statements are correct? Tick **one** box. [1]

   1. A habitat dominated by one species must have a high species richness.

   2. Species evenness takes into account the number of species present.

   3. An ecosystem with high species richness must have a low species evenness.

   1, 2 and 3 ☐   1 and 3 ☐   3 only ☐   None of the above ☐

2 Some scientists measured the areas of three farmers' fields. They then recorded the age of the fields.

The results are shown in the table.

Field	Area of field (m²)	Age of field (years)
A	4000	217
B	1500	250
C	10 000	162

a) Suggest a reason for the pattern shown between the age of a field and the area of the field. [2]

   ....

   ....

b) This formula can be used to estimate the number of plant species (per metre of hedge) using the age of the hedge:

$$\text{number of plant species per m of hedge} = \frac{\text{age of hedge in years} - 30}{110}$$

   Estimate the number of plant species per m of hedge for each of the fields in the table. [3]

   A = ...

   B = ...

   C = ...

c) Explain why conservationists are concerned about farmers destroying hedges on old fields to make larger fields. [3]

   ....

   ....

   ....

3 A scientist collected samples of water from two ponds, A and B.

The diagram shows the animals that were in the sample of water from pond A.

Pond A

a) Use the diagram to complete this table. [1]

Number of organisms of each species (*n*)				

b) Calculate the species diversity index for pond A. [3]

   ....

c) The species diversity index for pond B is 5.40

   Explain what these results indicate about which pond has the most stable community. [2]

   ....

Investigating diversity

Key words

Key questions

Why is it important to sample at random?

What does the standard deviation of a sample indicate?

Why is the amino acid sequence equivalent to the DNA base sequence?

Measuring species diversity

It is not possible to record all the organisms living in a habitat. Instead, sampling is used to estimate the true population of each organism.

There are two types of sampling: random and non-random. To estimate population size, random sampling is used. (Non-random sampling is used to find out about the distribution of organisms.)

Random sampling is used to eliminate sampling bias (where certain areas are sampled because they look interesting or easier to count).

The area being sampled is mapped out and the sampling areas are decided by random numbers to act as grid references. A quadrat is placed at each point and readings taken.

The measurements taken will vary depending on the investigation. They could include:

- the number of individuals of one or more species
- the percentage cover of one or more species
- quantitative data of a characteristic of one species to show variation in that species.

Once the data is collected, it is processed. A **mean** is usually calculated using readings from a number of quadrats. The **standard deviation** of the means can also be calculated to give an idea of the variation between the different samples.

Statistical tests can be used to see if there is a significant difference between the species diversity in two areas or to show variation within a species.

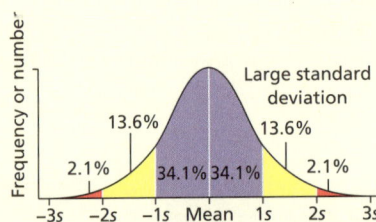

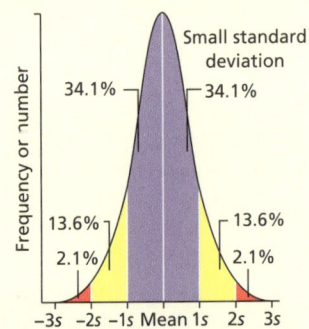

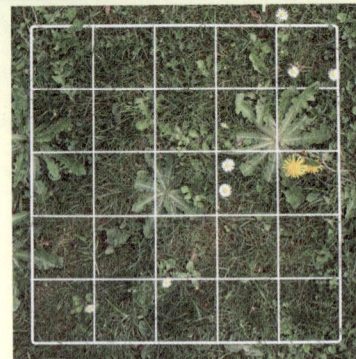

Investigating genetic diversity

Genetic diversity within, or between, species can be made by comparing the frequency of measurable or observable characteristics. Until recently, this was the only way of inferring genetic differences. Modern methods of **gene technology** now make more direct measurements of genetic variation possible. This has been used in taxonomy for phylogenetics (see page 74).

Comparisons can now be made between individuals within or between species using:

- the base sequence of DNA
- the base sequence of mRNA
- the amino acid sequence of the proteins encoded by DNA and mRNA.

These characteristics are all interrelated, as the base sequence of DNA determines the sequence in mRNA, which in turn determines the amino acid sequences of proteins.

It is now possible to make multiple copies of DNA from single molecules (see page 136). This means that genetic differences can be identified from very small samples.

Comparison of DNA sequences is used to identify individuals using genetic fingerprinting (page 140) and to screen and treat patients for various health risks (page 138).

Summary

Investigating diversity

(1) Which of these statements are correct? Tick **one** box. [1]

 1. There is no genetic diversity within a species.

 2. To measure species diversity in a habitat, non-random quadrats are used.

 3. Genetic diversity can be assessed by looking at mRNA base sequences.

  1, 2 and 3 ☐   1 and 3 ☐   3 only ☐   None of the above ☐

(2) Limpets are animals that live on the seashore. They have a large sucker under the shell that attaches them to rocks. Scientists measured the shell height and shell diameter of 12 limpets on a sheltered shore and 20 limpets on a shore exposed to heavy seas. The results are shown in the table.

Sheltered shore		Exposed shore	
Diameter (mm)	Height (mm)	Diameter (mm)	Height (mm)
44.3	34.2	61.0	26.7
50.0	30.5	64.5	22.4
46.8	33.9	57.0	23.0
47.5	34.5	63.0	29.0
46.9	30.2	55.8	23.0
50.0	26.9	61.0	14.0
48.4	34.1	52.2	18.3
48.0	22.7	58.3	26.0
35.5	30.8	57.1	13.0
50.0	25.0	52.0	19.0
52.0	25.8	55.4	19.0
44.6	30.5	53.9	26.0

Mean = 47.0   29.9

SD = 4.3   4.0   4.1   5.0

a) The scientists sampled the limpets at random on each shore. Give the reason why. [1]

b) Calculate the mean values for the shell diameter and height on the exposed shore. [2]

Shell diameter = ...

Shell height = ...

c) State what the standard deviation (SD) values indicate about the height of the shell on the exposed shore compared to the sheltered shore. [1]

d) Suggest an explanation for the differences between the mean values for shell height and diameter on the two different shores. [3]

(3) Scientists investigated the amino acid sequence in a section of haemoglobin from four mammals. The section was seven amino acids long. The table shows the scientists' results.

Mammal	Amino acid found at each location (the letters represent different amino acids)						
	1	2	3	4	5	6	7
Chimpanzee	N	T	R	P	A	E	H
Gibbon	D	K	R	Q	T	D	H
Gorilla	N	T	K	P	A	D	L
Oranqutan	N	K	R	Q	T	D	L

a) Use evidence from the table to explain which two mammals are:

 i) most closely related [2]

 ii) most distantly related [2]

b) Explain why studying the DNA base sequence coding for a protein may provide more information about relationships than studying the amino acid sequence of the protein. [2]

Photosynthesis: light-dependent reactions

🔑 Key words

❓ Key questions

What happens to chlorophyll a when it absorbs light?

What is photolysis?

What is the role of ATP synthase?

Light absorption

Photosynthesis takes place inside **chloroplasts**. The light-dependent reactions are on the grana and the light-independent reactions are in the stroma (see page 30).

Outer membrane — Inner membrane — Stroma thylakoid — DNA — Stroma — Thylakoid lumen — Ribosome — Granum — Thylakoid — Starch grain

On the thylakoids of the grana, are a range of pigments, the most important being **chlorophyll a**. This absorbs blue wavelengths (425–450 nm) and red wavelengths (600–700 nm), reflecting green (500–550 nm). The pigments are contained in two photosystems:

- The photosystems combine light-harvesting pigments with other proteins, to transfer excitation energy and electrons.
- Photons of light excite chlorophyll a in photosystem II (PSII) and cause it to lose electrons (**photoionisation**).
- This excitation drives the **photolysis** of water to form oxygen gas, hydrogen ions (protons) and electrons. The electrons from the water replace those lost by PSII.
- The electrons lost from PSII pass through a series of carriers in the **electron transfer chain** to reach photosystem I (PSI) and replace electrons that have been excited and lost from PSI.
- The electrons from PSI pass through more carriers and combine with the hydrogen ions from the water to reduce NADP to $NADPH_2$.

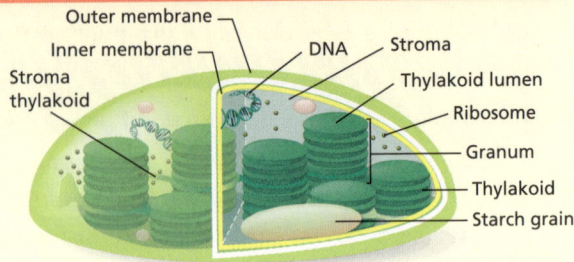

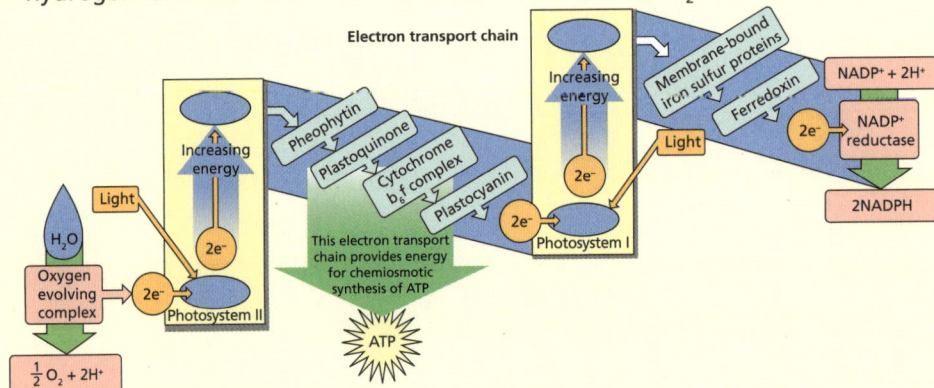

Electron transport chain

Pheophytin — Plastoquinone — Cytochrome b_6f complex — Plastocyanin — Increasing energy — Membrane-bound iron sulfur proteins — Ferredoxin — $NADP^+ + 2H^+$ — $NADP^+$ reductase — 2NADPH — Light — $2e^-$

H_2O — Oxygen evolving complex — Photosystem II — $2e^-$ — Light — Increasing energy — This electron transport chain provides energy for chemiosmotic synthesis of ATP — ATP — Photosystem I

$\frac{1}{2} O_2 + 2H^+$

ATP production

The movement of electrons through the electron transfer chain in the membranes of the thylakoids pumps hydrogen ions (protons) into the thylakoid from the stroma. This sets up a pH/proton gradient across the membrane.

The protons pass through ATP synthase back to the stroma. This causes ADP and inorganic phosphate ions (P_i) to form ATP.

This production of ATP is called **photophosphorylation** and the mechanism is called **chemiosmosis**.

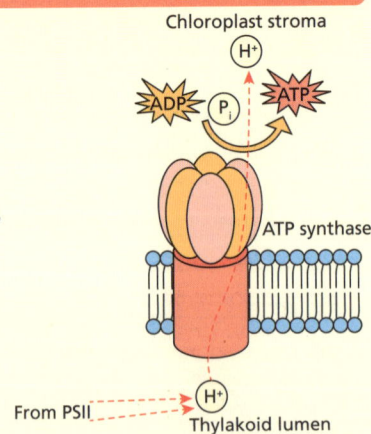

Chloroplast stroma — H^+ — ADP — P_i — ATP — ATP synthase — From PSII — H^+ — Thylakoid lumen

✔ Summary

Photosynthesis: light-dependent reactions

1. Which substances are made by the light-independent reactions of photosynthesis? Tick **one** box. [1]

ATP, $NADPH_2$ and O_2 ☐　　　　ATP, NADP and O_2 ☐

ADP, $NADPH_2$, O_2 and H_2O ☐　　　　ADP, NADP and H_2O ☐

2. Which of these statements are correct? Tick **one** box. [1]

   1. Electrons from PSI pass to PSII.

   2. The energy to phosphorylate ADP is stored in a proton gradient.

   3. The protons used to form $NADPH_2$ come from PSII.

   1, 2 and 3 ☐　　　　1 and 3 ☐　　　　2 only ☐　　　　None of the above ☐

3. Which molecule is the final electron acceptor in the light-dependent reactions? Tick **one** box. [1]

   Chlorophyll a ☐　　　　NADP ☐　　　　Oxygen ☐　　　　Water ☐

4. The diagram shows a chloroplast.

Give the letter from the diagram that labels each of these parts of the chloroplast. Each letter can be used once, more than once or not at all.

a) The site of the light-independent reactions

.... [1]

b) The position of chlorophyll a molecules

.... [1]

c) The area containing ATP synthase

.... [1]

d) An area that becomes **less** acidic as a result of the actions of the electron transport chain

.... [1]

5. Scientists investigated the effect of light on the pH of the stroma in chloroplasts. They measured the pH of the stroma of isolated chloroplasts for 12 minutes under conditions of dark, light and then dark. Their results are shown in the graph.

Use the light-dependent reactions to explain the scientists' results. [4]

....
....
....
....
....
....
....
....
....
....
....
....
....
....

Photosynthesis: light-independent reactions

Key words

Key questions

What is the role of RuBP in photosynthesis?

How many molecules of carbon dioxide are fixed to make one glucose molecule?

On a warm, bright day, what is usually limiting the rate of photosynthesis?

Carbon dioxide fixation

The light-independent reactions use reduced NADP from the light-dependent reactions to form a simple sugar. The hydrolysis of ATP, also from the light-dependent reaction, provides the additional energy for this reaction.

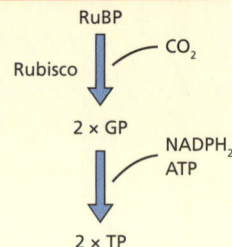

The first stage is carbon fixation:

- Carbon dioxide reacts with a five-carbon molecule called **ribulose bisphosphate (RuBP)** to form two three-carbon molecules of **glycerate 3-phosphate** (GP).
- This reaction is catalysed by the enzyme **ribulose bisphosphate decarboxylase (rubisco)**.
- ATP and reduced NADP from the light-dependent reaction are then used to reduce GP to the three-carbon sugar **triose phosphate (TP)**.

These reactions take place in the stroma.

RuBP → (Rubisco, CO_2) → 2 × GP → ($NADPH_2$, ATP) → 2 × TP

Calvin cycle

One in six of the triose phosphate molecules that are produced are converted to useful organic substances. These are initially sugars but from these sugars other substances (such as amino acids and lipids) are made.

Most of the molecules of TP are used to regenerate RuBP. This is done via a series of enzyme-catalysed reactions called the **Calvin cycle**.

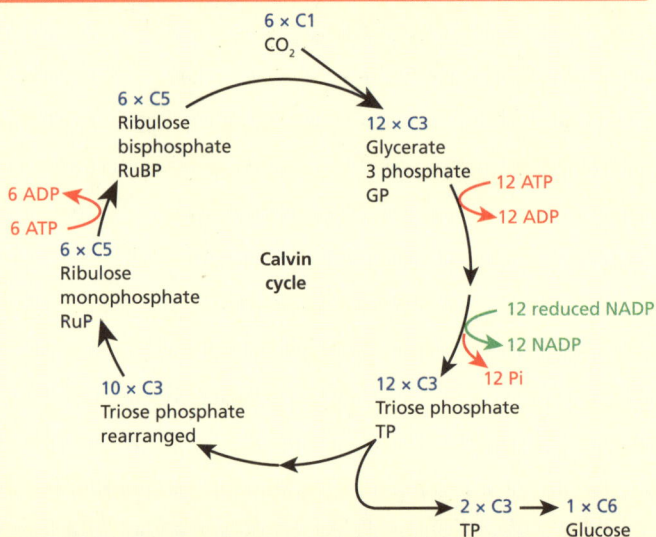

6 × C1 CO_2

6 × C5 Ribulose bisphosphate RuBP

12 × C3 Glycerate 3 phosphate GP

6 ADP
6 ATP

6 × C5 Ribulose monophosphate RuP

Calvin cycle

12 ATP
12 ADP

12 reduced NADP
12 NADP
12 Pi

10 × C3 Triose phosphate rearranged

12 × C3 Triose phosphate TP

2 × C3 TP → 1 × C6 Glucose

Limiting factors

If a component is in low supply then the productivity of photosynthesis is prevented from reaching its maximum. The feature that is preventing the reaction from going at a faster rate is called the **limiting factor**.

In photosynthesis, temperature, carbon dioxide and light intensity are key limiting factors:

- Temperature – affects the rate because it affects the enzymes that catalyse the steps in the light-independent reaction.
- Carbon dioxide – the enzyme rubisco has a low affinity to carbon dioxide, so at the usual atmospheric level of 0.04% carbon dioxide is often the limiting factor.
- Light intensity – can limit the rate if there are insufficient photons of light to excite the two photosystems.

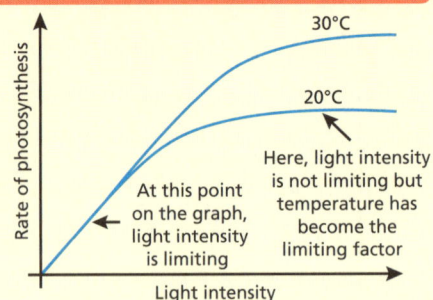

Rate of photosynthesis (y-axis), Light intensity (x-axis). 30°C, 20°C. At this point on the graph, light intensity is limiting. Here, light intensity is not limiting but temperature has become the limiting factor.

Summary

Spec. ref. 3.5.1

Photosynthesis: light-independent reactions

(1) Where are the enzymes that catalyse the Calvin cycle found in a plant cell? Tick **one** box. [1]

Attached to the chloroplast envelope ☐   Attached to the granal membranes ☐

In the stroma ☐   Inside the thylakoids ☐

(2) Which of these statements about the light-independent reactions are correct? Tick **one** box. [1]

1. GP accepts carbon dioxide molecules and is converted to TP.

2. $NADPH_2$ is reduced in the reactions and GP is oxidised.

3. The Calvin cycle regenerates TP.

1, 2 and 3 ☐   1 and 3 ☐   2 only ☐   None of the above ☐

(3) An experiment was carried out to investigate the effect of reducing the carbon dioxide concentration available to a plant. The concentration of RuBP and GP in the plant were measured. The results are shown on the graph.

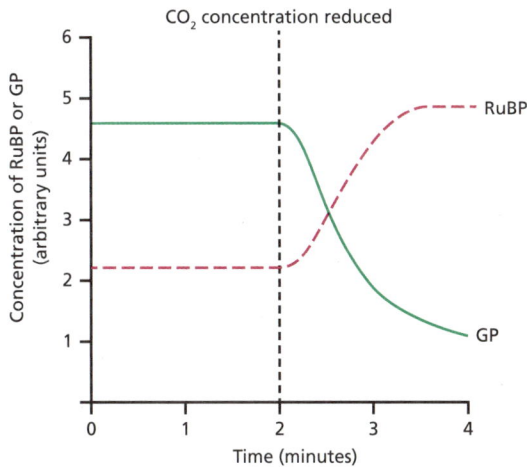

a) Explain the effects of reducing the carbon dioxide concentration on:

i) RuBP concentration [2]

....

....

....

....

ii) GP concentration [2]

....

....

....

....

b) Explain how the concentration of TP would change over the same time period. [2]

....

....

(4) In most plants, rubisco is inefficient in fixing carbon dioxide because oxygen competes with the carbon dioxide for its active site. The graph shows how the fixation of carbon dioxide and oxygen changes with temperature.

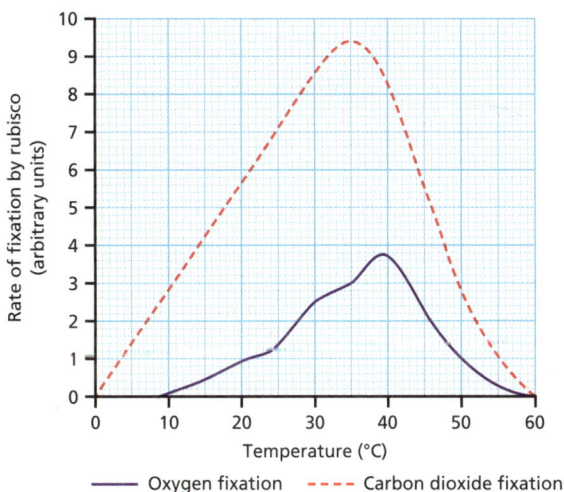

Oxygen fixation ——   Carbon dioxide fixation - - -

a) Compare the fixation of carbon dioxide with the fixation of oxygen, as shown in the graph. [3]

....

....

....

....

....

b) Some plants exist that use a different enzyme for carbon dioxide fixation. However, the reactions of photosynthesis in these plants need more energy.

Explain why these plants outcompete normal plants in hot, sunny climates but are outcompeted by normal plants in less sunny climates. [4]

....

....

....

Glycolysis and anaerobic respiration

🔖 **Key words**

Key questions

What is the net ATP production of glycolysis?

How many molecules of NADH$_2$ are made in glycolysis?

Why is it important to regenerate NAD?

Glycolysis

Respiration is a catabolic process, which means that large molecules, such as glucose, are broken down into smaller ones.

The first stage in respiration is **glycolysis**:

- This occurs in the cytoplasm of the cell.
- The first step is for the six-carbon glucose molecule to be phosphorylated from the conversion of ATP into ADP.
- This creates **glucose phosphate** and then fructose bisphosphate.
- This molecule is broken (lysed) into two **triose phosphate** molecules.
- Triose phosphate is then oxidised to form two molecules of **pyruvate**, via the conversion of four molecules of ADP into four molecules of ATP and two molecules of NAD being reduced to form two molecules of NADH$_2$.

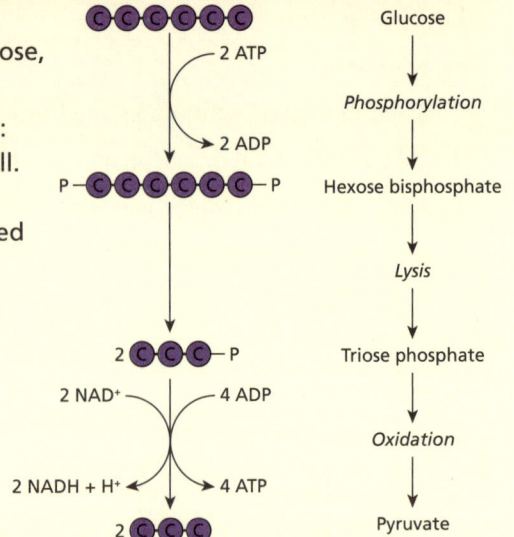

This means that, although two molecules of ATP have to be used at the start of glycolysis, four ATP are made at the end. Therefore, there is a net production of two ATP and two NADH$_2$ molecules by glycolysis. Glycolysis does not involve the mitochondria, or any oxygen – it is therefore anaerobic.

Glucose
↓
Phosphorylation
↓
Hexose bisphosphate
↓
Lysis
↓
Triose phosphate
↓
Oxidation
↓
Pyruvate

Anaerobic respiration

At the end of glycolysis, the next step in respiration depends on oxygen availability. If oxygen is limited then the cell must follow an anaerobic pathway.

In animal cells, pyruvate is reduced to **lactate** by the enzyme **lactate dehydrogenase**. This requires NADH$_2$, which is oxidised to NAD.

The regeneration of NAD is important because if it did not occur, the cell would run low on NAD. That would prevent glycolysis from happening.

Anaerobic respiration is common in active muscle and the lactate produced causes muscle cramps and fatigue. After exercise, the lactate can be converted back to pyruvate or glycogen in the liver.

In organisms such as yeast, pyruvate is not converted to lactate:

- It is first converted to **ethanal** and carbon dioxide.
- The ethanal is then reduced to **ethanol** by the enzyme **ethanol dehydrogenase**.
- This second step causes the regeneration of NAD.

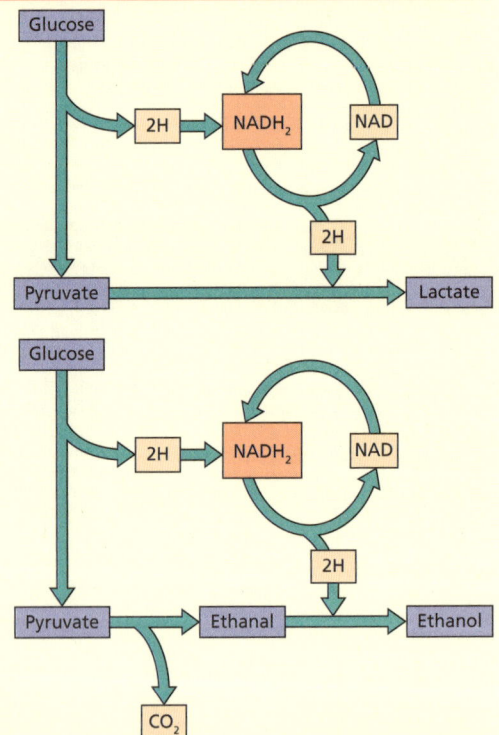

✔ **Summary**

Glycolysis and anaerobic respiration

1 What are the net products of glycolysis from one molecule of glucose? Tick **one** box. [1]

1 ATP, 1 $NADH_2$ and 1 pyruvate ☐   2 ATP, 2 $NADH_2$ and 2 lactate ☐

2 ATP, 2 $NADH_2$ and 2 pyruvate ☐   4 ATP, 2 $NADH_2$ and 2 pyruvate ☐

2 Which of these statements about anaerobic respiration are correct? Tick **one** box. [1]

1. The conversion of pyruvate to ethanal involves decarboxylation.

2. NADP is regenerated by the reduction of pyruvate.

3. The only ATP produced in anaerobic respiration is from glycolysis.

1, 2 and 3 ☐   1 and 3 only ☐   2 and 3 only ☐   1 only ☐

3 The diagram shows reactions occurring in a yeast cell as part of anaerobic respiration.

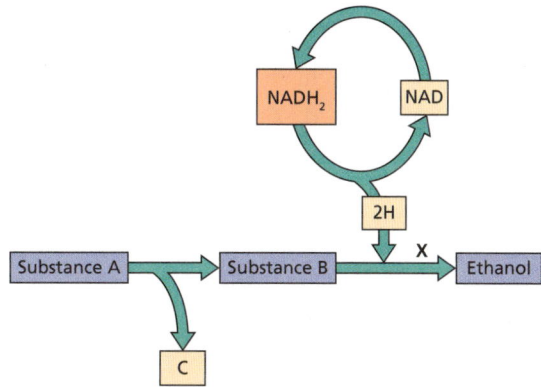

a) Identify substances A, B, C. [3]

A =

B =

C =

b) Name the enzyme that catalyses reaction X. [1]

c) Give **two** ways that anaerobic respiration differs in a mammalian cell. [2]

4 The diagram shows the five main steps in glycolysis.

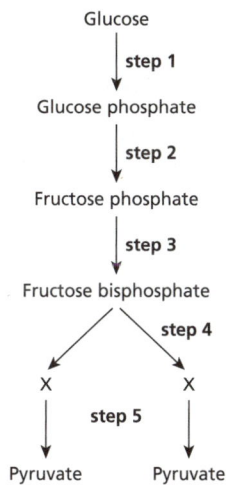

Glucose
step 1
Glucose phosphate
step 2
Fructose phosphate
step 3
Fructose bisphosphate
step 4
X — X
step 5
Pyruvate — Pyruvate

a) Where in the cell does glycolysis occur? [1]

b) Name substance X. [1]

c) Give the number of the step or steps in which each of these processes occur. [4]

Lysis

Phosphorylation

Production of ATP

Production of $NADH_2$

Aerobic respiration

Key words

Key questions

How many molecules of acetyl CoA are made from one glucose molecule?

The link reaction

If oxygen is available in the cell, pyruvate that is made by glycolysis is transported into the mitochondria.

In the matrix, the pyruvate is decarboxylated with **coenzyme A** (CoA) by the removal of CO_2 to form **acetyl coenzyme A** (acetyl CoA).

$$Pyruvate + CoA \xrightarrow{\quad NAD \quad NADH_2 \quad} acetyl\ CoA + CO_2$$

The process requires the reduction of NAD to form $NADH_2$.

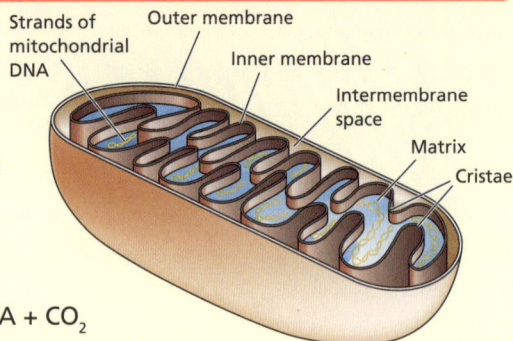

Labels on mitochondrion diagram: Strands of mitochondrial DNA; Outer membrane; Inner membrane; Intermembrane space; Matrix; Cristae

The Krebs cycle

The Krebs cycle occurs in the matrix and converts acetyl CoA into different organic molecules:

- The first step is the formation of **citrate** from the reaction of **oxaloacetate** with acetyl CoA.
- Citrate is then converted into other molecules, producing two CO_2 molecules from the decarboxylation of the intermediate molecules.
- A total of three $NADH_2$ molecules and one $FADH_2$ molecule are produced for every turn of the cycle.
- One ATP molecule is also produced per turn.
- The cycle regenerates oxaloacetate to combine with more acetyl CoA.

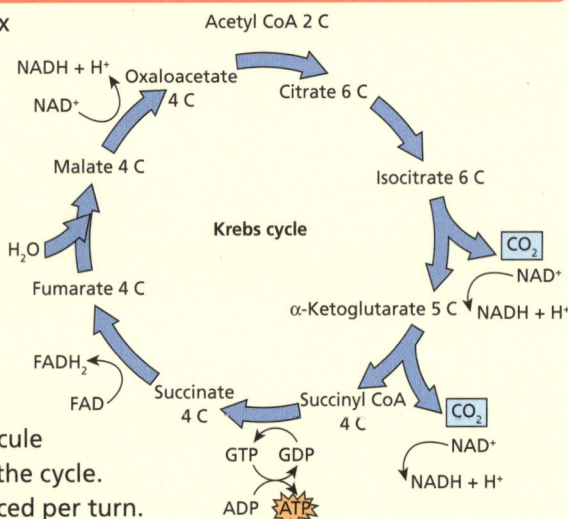

Krebs cycle diagram labels: Acetyl CoA 2 C; $NADH + H^+$; NAD^+; Oxaloacetate 4 C; Citrate 6 C; Malate 4 C; Isocitrate 6 C; H_2O; Krebs cycle; Fumarate 4 C; CO_2; NAD^+; $FADH_2$; FAD; Succinate 4 C; Succinyl CoA 4 C; α-Ketoglutarate 5 C; $NADH + H^+$; GTP; GDP; CO_2; NAD^+; $NADH + H^+$; ADP; ATP

Why is the Krebs cycle described as oxidative decarboxylation?

Although oxygen is not directly used in the Krebs cycle, without regeneration of NAD and FAD in oxidative phosphorylation, the Krebs cycle would stop.

Oxidative phosphorylation

The final stage of respiration is **oxidative phosphorylation**. This occurs by chemiosmosis (see page 80). It uses the **electron transport chain**, located on the inner membranes of the mitochondria.

- Electrons from $NADH_2$ or $FADH_2$ pass along the carriers.
- This pumps proteins from the matrix into the intermembrane space, setting up a proton gradient.
- Protons are then allowed to re-enter the matrix through ATP synthase.
- This is coupled with the phosphorylation of ADP to form ATP.
- Oxygen accepts the electrons and the protons to form water.

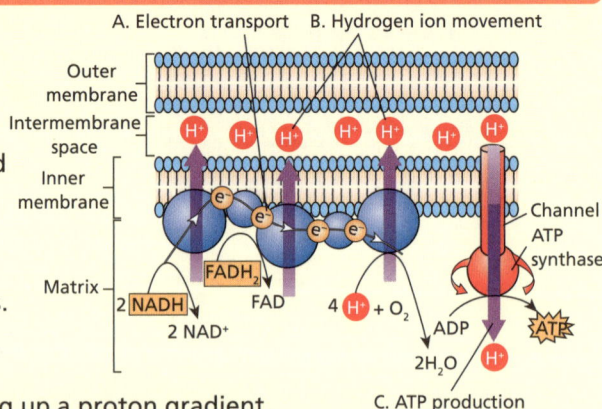

How does the movement of protons differ in mitochondria compared to in thylakoids?

Oxidative phosphorylation diagram labels: A. Electron transport; B. Hydrogen ion movement; Outer membrane; Intermembrane space; Inner membrane; Matrix; H^+; e^-; 2 NADH; $FADH_2$; FAD; 2 NAD^+; 4 H^+ + O_2; ADP; $2H_2O$; H^+; Channel ATP synthase; ATP; C. ATP production

Summary

Aerobic respiration

(1) Which of these statements describes the link reaction? Tick **one** box. [1]

Pyruvate is oxidised and decarboxylated. ☐   Pyruvate is reduced and decarboxylated. ☐

Acetyl CoA is oxidised and decarboxylated. ☐   Acetyl CoA is reduced and decarboxylated. ☐

(2) Which of these statements about aerobic respiration are correct? Tick **one** box. [1]

1. Glycolysis is not necessary for the Krebs cycle to continue.
2. $NADPH_2$ is generated in the link reaction.
3. The Krebs cycle occurs twice for each glucose molecule respired.

1, 2 and 3 ☐   2 and 3 only ☐   1 and 2 only ☐   3 only ☐

(3) The table shows the number of ATP molecules produced in different stages of respiration.

Stage	ATP produced directly	Coenzyme production	ATP produced by oxidative phosphorylation	Total ATP production
Glycolysis	2ATP	2 $NADH_2$	4–6 ATP	6–8 ATP
Link reaction	None	2 $NADH_2$	6 ATP	6 ATP
Krebs cycle		6 $NADH_2$ 2 $FADH_2$	18 ATP 4 ATP	24 ATP

a) Complete the table by writing the number of ATP molecules produced directly by the Krebs cycle and the total range of ATP molecules produced by respiration. [2]

b) ATP production from $NADH_2$ made in glycolysis is lower than $NADH_2$ made from processes occurring in the mitochondria. That is because ATP is required to transport $NADH_2$ into mitochondria.

Explain why ATP is needed for this process. [2]

c) The electrons from $FADH_2$ feed into the electron transport chain part way along.

Explain how this can account for the difference between ATP production from $FADH_2$ compared with $NADH_2$ [2]

(4) Scientists investigate ATP production by chemiosmosis in mitochondria. They remove the outer membrane of mitochondria, as shown in the diagram. They then measure the ATP production by the mitochondria.

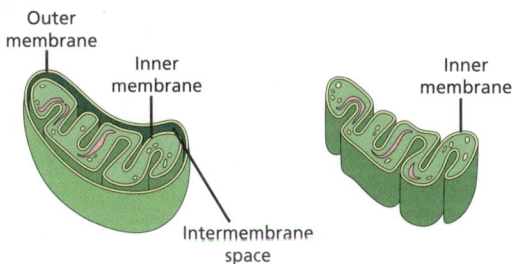

Outer membrane, Inner membrane, Intermembrane space, Inner membrane

a) Explain why ATP production is lower in these mitochondria. [4]

b) Explain why it is important that the inner phospholipid bilayer is impermeable to hydrogen ions. [2]

Energy flow in ecosystems

Key words

Key questions

Why is it often better to measure dry mass than wet mass?

Why is the value for R usually higher in animals than in plants?

Why are farmers interested in knowing productivity of crops?

Biomass

In ecosystems, plants synthesise organic compounds from carbon dioxide. Most of the sugars synthesised by plants are used by the plant in respiration. The rest are used to make other groups of biological molecules. These biological molecules form the **biomass** of the plants.

The biomass of plants can be measured as a mass of tissue per unit area. The biomass can be:

- wet biomass, which includes the water content of the cells
- dry biomass, which is harder to measure as it involves heating the material in an oven to remove the water content.

The chemical energy store in dry biomass can be estimated using a **calorimeter** (shown right). This burns the material and the energy transferred to the water can be calculated by measuring the increase in temperature.

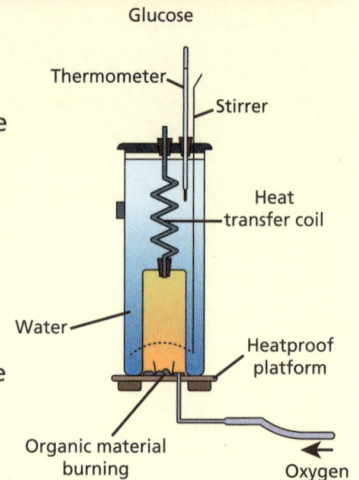

Glucose, Thermometer, Stirrer, Heat transfer coil, Water, Heatproof platform, Organic material burning, Oxygen

Productivity

Most of the sunlight energy falling on a plant is not absorbed and therefore not used in photosynthesis. Some is of the wrong wavelength.

Gross primary production (GPP) is the chemical energy that is used and trapped in plant biomass. It is usually quoted for a given area or volume. Much of this biomass is used in respiration and so the energy is transferred to the environment.

Net primary production (NPP) is the chemical energy store in plant biomass after respiratory losses to the environment have been deducted. Therefore:

$$NPP = GPP - R$$ (where GPP represents gross production and R represents respiratory losses to the environment.)

99% of solar energy is reflected or passes through producers without being absorbed

1% of solar energy striking producers is captured by photosynthesis (GPP)

GPP

60% of GPP is lost to respiration

NPP

40% of GPP supports the growth and reproduction of producers (NPP)

This NPP is available for plant growth and reproduction. It is also available to other trophic levels in the ecosystem, such as herbivores and decomposers.

The **net production of consumers (N)**, such as animals, can be calculated as:

$$N = I - (F + R)$$ (where I represents the chemical energy store in ingested food; F represents the chemical energy lost to the environment in faeces and urine; and R represents the respiratory losses to the environment.)

Calculating the NPP or N over a certain length of time gives a measure of **productivity**. Primary or secondary productivity is the rate of primary or secondary production, respectively. It is measured as biomass in a given area, in a given time, e.g. $kJ\ ha^{-1}\ year^{-1}$.

Productivity is affected by **efficiency** of energy transfer between a food chain's trophic levels.

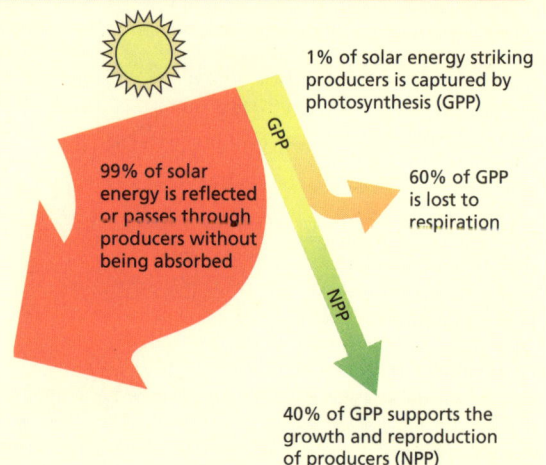

Summary

Energy flow in ecosystems

1. A calorimeter, like the one shown on page 88, is used to measure the energy content of a sample of biomass. There is 20 g of dry biomass at the start of the experiment and 200 cm^3 (200 g) of water in the apparatus. The temperature on the thermometer rose from 20°C to 35°C.

   a) Give **two** ways that the apparatus is designed to make sure that as much energy as possible is transferred to the water and not lost to the air. [2]

   b) State how the apparatus is adapted to give complete combustion of the sample. [1]

   c) The formula used to calculate the energy needed to increase the temperature of the water is:

   energy needed (J) = mass of water (g) × temperature change (°C) × 4.2

   Use this formula to calculate the energy content of the biomass sample in J/g. [3]

2. The diagram shows a simple food chain.

   Not to scale

   Grass → Grasshopper → Shrew → Owl

   The NPP of the grass is 6.5×10^4 kJ m^{-2}. Energy transfer from producer to consumer is about 10% and from producer to producer it is about 20%.

   a) Calculate a value for the net production of the owl. [2]

   b) Explain why food chains rarely have more than four or five trophic levels. [2]

   c) Suggest why the efficiency of transfer from consumer to consumer is more efficient than from producer to consumer. [2]

3. The diagram shows the energy flow and productivity of organisms in a food chain.

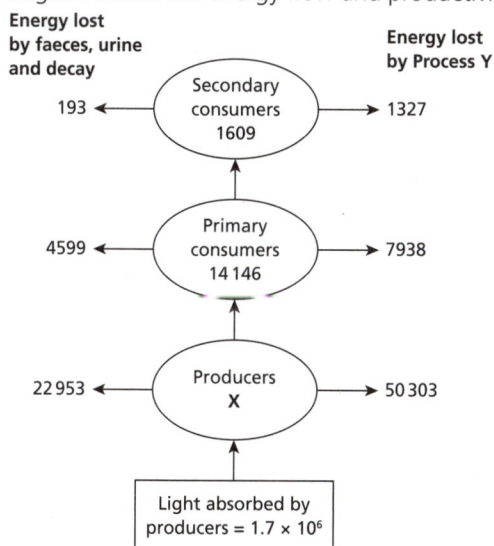

Energy lost by faeces, urine and decay

Energy lost by Process Y

Secondary consumers 1609
193 ←   → 1327

Primary consumers 14 146
4599 ←   → 7938

Producers X
22 953 ←   → 50 303

Light absorbed by producers = 1.7×10^6

   a) Suggest units for the energy values shown in the diagram. [1]

   b) i) Calculate the value X for the producers. [1]

   ii) Give the name given to this value. [1]

   c) Name the process Y on the diagram. [1]

   d) Calculate how much energy would be available for tertiary consumers. [1]

   e) The value X on the diagram is much lower than the value for the energy of light absorbed by the producers.

   Give a reason why some of the energy absorbed by the producers cannot be used in photosynthesis. [1]

Nutrient cycles

Key words

Key questions

In mycorrhizae, what do the fungi gain from the plant?

Which bacteria in the nitrogen cycle are aerobic?

Why are artificial fertilisers fast-acting?

Nutrient cycles in nature

The biomass of dead or decaying organisms is broken down by decomposers (**saprobionts**). These organisms are usually bacteria and fungi. They make the elements in organic material, such as nitrogen and phosphorus, available for other organisms to utilise. The nitrogen and phosphorus that make up an organism will have been part of many other organisms and are recycled repeatedly.

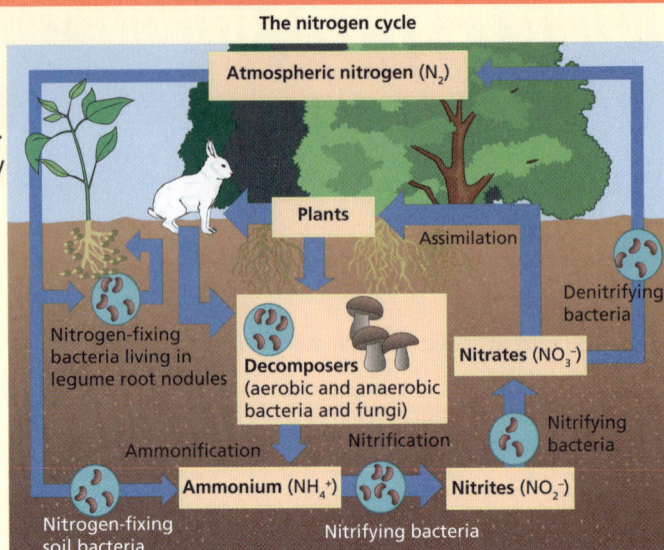

The nitrogen cycle

- Atmospheric nitrogen (N_2)
- Plants
- Assimilation
- Denitrifying bacteria
- Nitrogen-fixing bacteria living in legume root nodules
- Decomposers (aerobic and anaerobic bacteria and fungi)
- Nitrates (NO_3^-)
- Nitrifying bacteria
- Ammonification
- Nitrification
- Nitrogen-fixing soil bacteria
- Ammonium (NH_4^+)
- Nitrites (NO_2^-)
- Nitrifying bacteria

As well as saprobionts, there are other types of bacteria involved in the nitrogen cycle:

- In **ammonification**, saprobiontic organisms decompose dead organic material to release ammonium ions.
- **Nitrification** involves aerobic **nitrifying bacteria** converting the ammonium ions to nitrites and then nitrates.
- Nitrates can then be absorbed by plant roots, often assisted by **mycorrhizae**, which are a mutualistic relationship between plant roots and fungi in the soil.
- Some of the nitrates in the soil are converted into nitrogen gas by **denitrification** carried out by anaerobic **denitrifying bacteria**.
- **Nitrogen-fixing bacteria**, in the soil or in plant root nodules, fix nitrogen into ammonium ions.

Nutrient cycles and farming

For many centuries, farmers have added natural fertilisers to their fields to replace the minerals that have been taken up by their crops. These natural fertilisers, such as compost and manure, contain organic material that is then decomposed in the soil.

- Abundance of nutrients causes algae bloom
- Lack of oxygen suffocates fish
- Waste and fertilizer runoff and leaching
- Nitrates Phosphates
- Algae blocks sunlight and die
- Bacteria use up oxygen to break down decaying matter
- Decay

Over the last 100 years, artificial fertilisers have become available. These are chemically-produced substances, such as ammonium phosphates and nitrates. Plants can take up these chemicals directly.

Use of artificial fertilisers has increased yields but issues have occurred by their overuse:

- Artificial fertilisers are very soluble and easily **run off** or **leach** into streams and lakes.
- The minerals in the fertilisers are taken up by plants in the water and stimulate their growth, causing a bloom.
- The algae die and saprobionts decompose them, using up all the oxygen in the water.
- Animals in the water then die from lack of oxygen.

This sequence of events is called **eutrophication**.

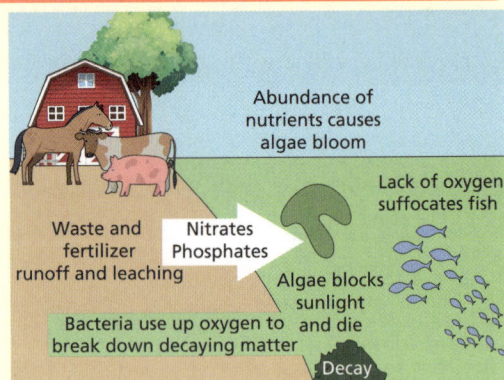

Summary

Nutrient cycles

1. Which of these statements about the nitrogen cycle is correct? Tick **one** box. [1]

In anaerobic conditions, nitrate ion concentration in the soil increases. ☐

Nitrogen fixation only occurs in the soil. ☐

Nitrification requires oxygen. ☐

Saprobionts produce nitrate ions. ☐

2. Scientists investigated the effect of mycorrhiza and artificial fertiliser on the growth of potatoes. They grew potatoes in three pots and treated each pot differently.

   - Pot 1 had no added treatment.
   - Pot 2 had chemical fertiliser added.
   - Pot 3 had mycorrhiza and organic fertiliser added.

   The table shows the results of their experiment.

	Dry mass of leaves in grams	Dry mass of potato tubers in grams
Pot 1	18	104
Pot 2	25	216
Pot 3	32	214

   a) Give the role of pot 1 in this investigation. [1]

   b) Calculate the percentage difference between:

   i) the dry mass of potato tubers in pot 1 and the dry mass of potato tubers in pot 2 [2]

   ii) the dry mass of potato tubers in pot 1 and the dry mass of potato tubers in pot 3 [2]

   c) Discuss the results of the investigation. [4]

   d) The scientists advised that farmers use the treatment given to pot 3 on their fields.
   Suggest why they gave this advice. [2]

3. Scientists want to prevent nitrates from artificial fertilisers passing into rivers. They set up a trial using woodchips coated with bacteria. The trial is shown in the diagram.

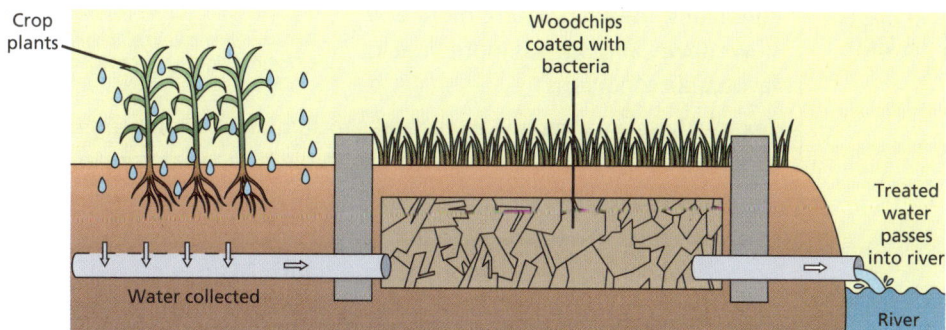

   a) Name the process that causes artificial fertilisers to pass through soil and into rivers. [1]

   b) The bacteria used to coat the woodchips is a type of bacteria from the nitrogen cycle.

   i) Name this type of bacteria. [1]

   ii) Explain why this type of bacteria is used. [2]

Response to stimuli

Key words

Key questions

What advantage does phototropism give to a plant?

What advantage does gravitropism give to a plant?

What is the advantage of simple reflexes not involving thought?

Responding to stimuli

Organisms increase their chance of survival by responding to changes in their environment. A change in the internal or external environment is a **stimulus**. This is detected by a **receptor**. A **coordinator** then formulates a suitable **response** and an **effector** carries out the response.

Responses in plants

Flowering plants produce specific growth factors in response to stimuli. These growth factors move to areas in the plant where they regulate growth. **Indoleacetic acid** (IAA) is one growth factor and is responsible for **tropisms**. Tropisms are directional growth responses made as a result of a stimulus. A growth response to light from a certain direction is a **phototropism** and a response to gravity is a **gravitropism**. These growth responses can be positive (towards) or negative (away).

- IAA is produced in the tip of a shoot and is transported down the shoot to cause cells to elongate.
- Unidirectional light causes more auxin to be sent to the shaded side of the shoot.
- Greater cell elongation on this side causes the shoot to grow towards the light, i.e. positive phototropism.

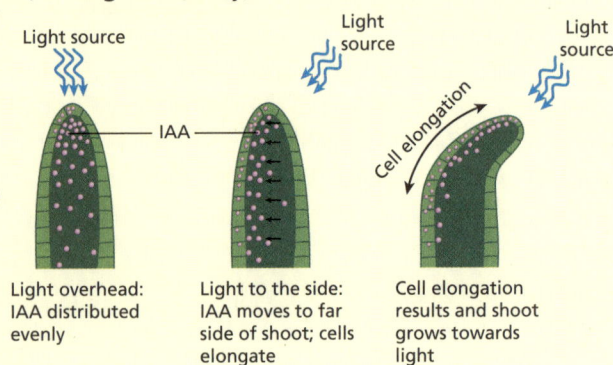

Light overhead: IAA distributed evenly

Light to the side: IAA moves to far side of shoot; cells elongate

Cell elongation results and shoot grows towards light

In roots, IAA has the opposite effect on cells. IAA is transported to the lower side of a root and inhibits cell elongation. This results in the root growing downwards towards gravity, i.e. positive gravitropism.

Responses in animals

Many responses in mammals are modified by experience and higher thinking skills. But simple responses exist in many animals. These include:

Kinesis – a non-directional response to a stimulus. When an animal detects that it is in an environment that is unfavourable, it increases its rate of movement. When in favourable conditions, the animal turns more and slows down. So it spends more time in favourable conditions.

The diagram shows a section through a choice chamber, often used for investigating responses of small animals. They experience different conditions and their movement can be observed.

Taxes – these are directional responses to stimuli. Taxic movements will move the organism directly away from an unfavourable stimulus or towards a favourable stimulus.

Simple reflexes – higher animals, such as mammals, still use simple reflexes to respond to some stimuli. They are rapid, do not involve conscious thought and protect the body from damage. A sensory neurone transmits an impulse from a receptor to the central nervous system (CNS). A relay neurone is then stimulated and in turn a motor neurone. This stimulates an effector to remove the animal from danger.

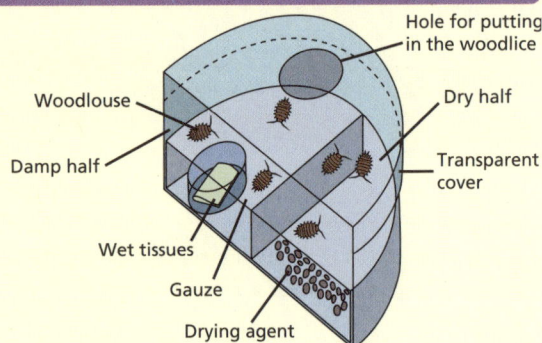

Summary

Response to stimuli

1. Give the order of types of neurones involved in a simple reflex. [1]

....

2. Students set up a choice chamber with water under one side of the chamber and calcium chloride under the other. Calcium chloride absorbs water from the air. The students put a woodlouse in the chamber and observed its movement. The diagram shows the movement of the woodlouse.

   a) Compare the movement of the woodlouse in each side of the chamber. [2]

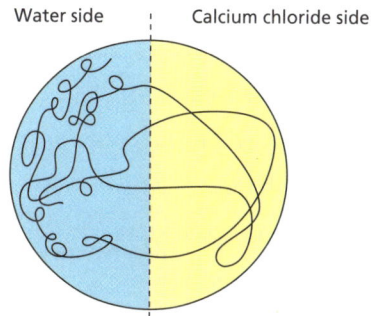

   Water side | Calcium chloride side

   ....

   ....

   b) Name the type of behaviour shown by the woodlouse. [1]

   ....

   c) Suggest an explanation for this behaviour. [3]

   ....

   ....

3. The diagrams show investigations into phototropism in shoots using four shoots. Mica is an impermeable substance.

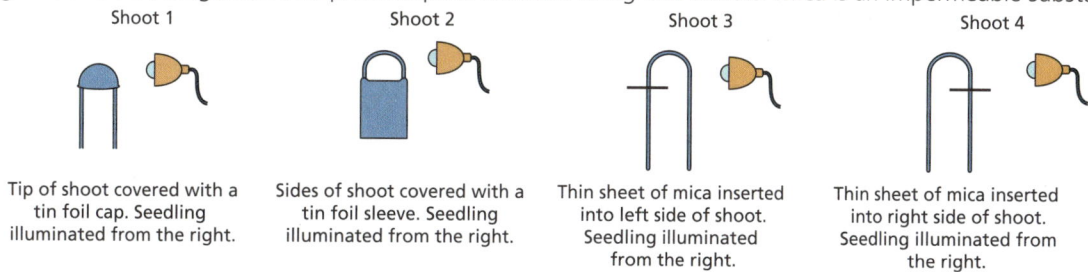

Shoot 1	Shoot 2	Shoot 3	Shoot 4
Tip of shoot covered with a tin foil cap. Seedling illuminated from the right.	Sides of shoot covered with a tin foil sleeve. Seedling illuminated from the right.	Thin sheet of mica inserted into left side of shoot. Seedling illuminated from the right.	Thin sheet of mica inserted into right side of shoot. Seedling illuminated from the right.

   Complete this table for each of the investigations. For the explanations, state the distribution of IAA. [8]

	Response of shoot	Explanation of response
Shoot 1		
Shoot 2		
Shoot 3		
Shoot 4		

4. The graph shows the effects of different concentrations of IAA on small pieces of roots and shoots.

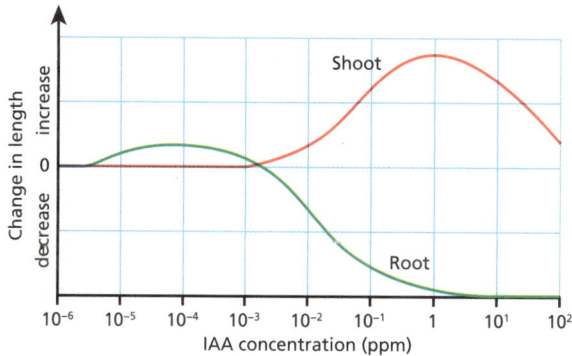

   a) The scale used on the x-axis is a logarithmic scale. Suggest why this type of scale is used. [1]

   ....

   ....

   b) Compare the effect of high concentrations of IAA on root and shoot growth. [2]

   ....

   ....

   ....

   c) IAA controls gravitropism in roots and phototropism in shoots.

   Explain how the graph shows that IAA concentration in intact roots and shoots must be above 10^{-3} ppm. [3]

   ....

   ....

   ....

Receptors

Key words

Key questions

What is the difference between a resting potential and a generator potential?

Why are Pacinian corpuscles found deep in the skin?

Which type of receptor is stimulated when looking straight at an object??

Pacinian corpuscles

Receptors are used in organisms to detect internal and external stimuli. In animals they share certain characteristics:

- They are sensitive to a specific stimulus, e.g. photoreceptors detect light, chemoreceptors detect chemicals and mechanoreceptors detect deformation.
- There is usually a potential difference across the cell membrane of the receptor, with the inside negative compared to the outside – this is called the **resting potential**.
- Stimulation of a receptor causes a change in the potential difference across the membrane – this is called a **generator potential**.

Pacinian corpuscles are mechanoreceptors. They are up to 2 mm long and are found deep in the skin. They have a capsule made of many layers of connective tissue around the end of the axon of a sensory neurone. Pacinian corpuscles detect vibrations and pressure on the skin, not light touch:

- These stimuli deform the lamellae, putting pressure on the end of the sensory neurone.
- This causes sodium ion channels in the membrane to open so sodium ions enter the axon.
- The outside of the axon becomes negatively charged compared to the inside, creating a generator potential.

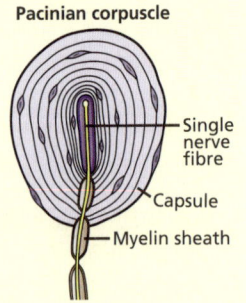

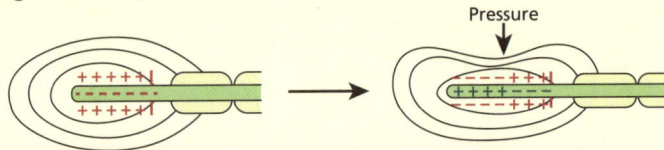

Pacinian corpuscle

Single nerve fibre
Capsule
Myelin sheath

Pressure

Rods and cones

The human retina of the eye contains two types of photoreceptors, called rods and cones.

Rod cells are:
- able to detect shades of light and dark, not colours
- able to operate in low light intensities
- spread over the periphery of the retina
- sensitive to light due to presence of the pigment rhodopsin, which sets up a generator potential when stimulated by light.

Cone cells:
- are mainly in the centre of the retina (fovea)
- can detect colour
- need high light intensities to work
- can detect information to give a highly detailed image (high visual acuity)
- are sensitive to colours because they have three photosensitive colour pigments (red, green and blue) that detect different colours/wavelengths of light.

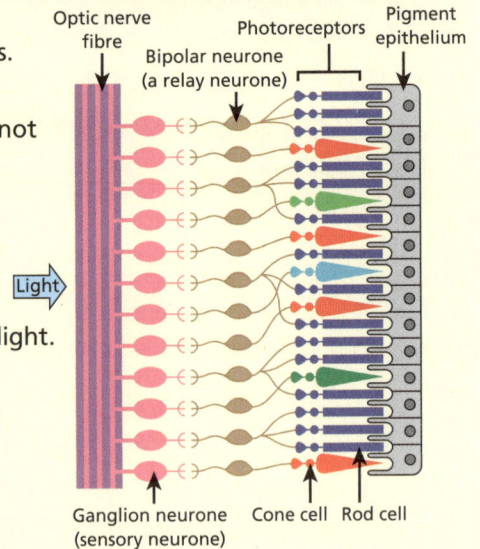

Differences in the sensitivity of rods and cones can be explained by their neural connections. Rods have many connections so are sensitive to dim light but have a low acuity. Cones have few connections so have a high acuity but need high light intensities.

Optic nerve fibre
Bipolar neurone (a relay neurone)
Photoreceptors
Pigment epithelium
Light
Ganglion neurone (sensory neurone)
Cone cell   Rod cell

Summary

Receptors

(1) Complete the table by adding ticks (✓) or crosses (✗) to show the features of rod and cone cells in the human retina. [5]

Feature	Rods	Cones
Produces a generator potential when stimulated by light		
Found mainly in the fovea		
Contains pigments sensitive to light		
Provides high visual acuity		
Are sensitive to most wavelengths of light		

(2) The two diagrams show:
- the structure of a Pacinian corpuscle from the skin of the fingertip
- the effect on the membrane potential of region Y when pressure is applied to the fingertip.

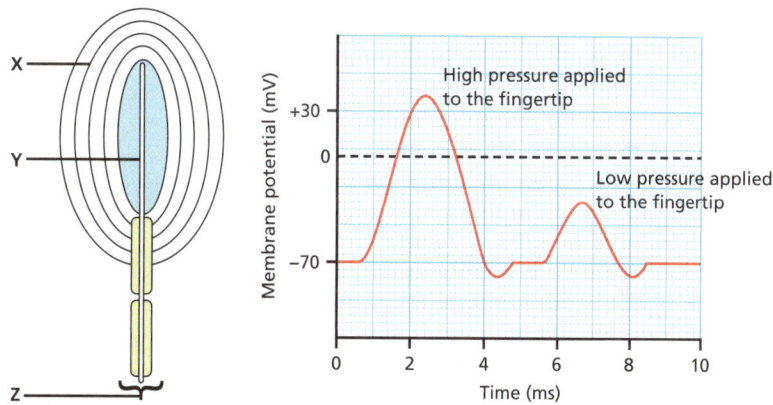

a) Identify regions X, Y and Z on the Pacinian corpuscle. [3]

X = ... Y = ... Z = ...

b) Give the name used to describe the membrane potential generated by pressure being applied to the skin. [1]

....

c) Explain how pressure on the skin produces the change in membrane potential. [3]

....

....

d) Describe the difference in the response of the Pacinian corpuscle to high and low pressure. [2]

....

....

(3) The diagram shows two types of sensory cell from the human retina.

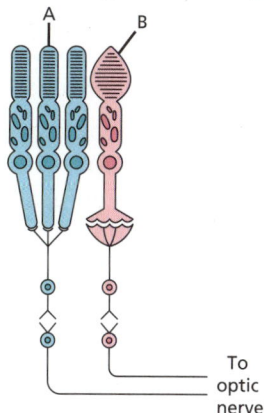

a) Name the types of cell labelled A and B. [1]

A = ...

B = ...

b) Explain why one type of cell gives greater visual acuity than the other type. [3]

....

....

....

....

c) Explain why one type of cell is more sensitive in dim light than the other type. [3]

....

....

Control of heart rate

🔖 Key words

❓ Key questions

Why is a natural pacemaker needed?

What is the function of the Purkinje fibres?

Why are the sympathetic and parasympathetic nerves described as antagonistic?

Coordinating the heart

Cardiac muscle cells each have their own inbuilt rhythm. Pacemaker areas of the heart coordinate the cardiac muscle to ensure that all the fibres contract at the same time:

- The main pacemaker is the **sinoatrial node** (SAN), which is in the wall of the right atrium.
- The SAN initiates a wave of electrical excitation across both atria. This causes the cardiac muscle in the atria to contract, the right atrium slightly before the left atrium.
- The wave of electrical energy reaches a second pacemaker, the **atrioventricular node** (AVN), which is located at the base of the right atrium.
- There is a ring of non-conductive tissue separating the atria from the ventricles, which means that the electrical excitation can only pass down the septum.
- The impulses pass in special fibres called **Purkinje fibres** in a group called the **bundle of His**.
- The excitation therefore spreads from the base of the heart up into the walls of the left and right ventricles, causing them to contract.

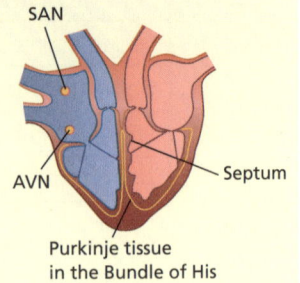

This pathway of excitation makes sure that the ventricles contract from the base upwards so that blood is efficiently ejected from the ventricles.

Controlling heart rate

The heart rate needs to change so that the body can correct variations in blood pressure and carbon dioxide content of the blood:

- Blood pressure is detected by pressure receptors (mechanoreceptors) in the carotid artery and aorta.
- The carbon dioxide content of the blood (actually pH) is detected by chemoreceptors in the carotid artery, aorta and in the **medulla oblongata** of the brain.

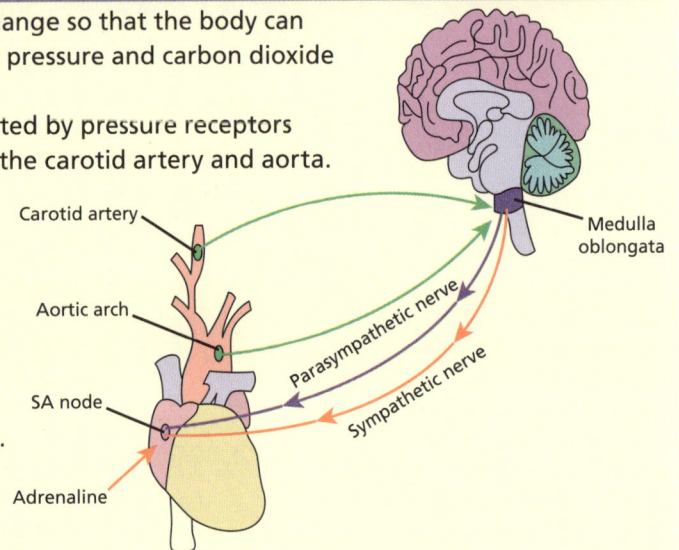

The **cardiac centre** of the medulla oblongata can alter the activity of the pacemaker. This changes the strength and rate of heartbeats. The **sympathetic nerve** stimulates an increase in heart rate, whilst the **parasympathetic nerve** (the **vagus**) causes a decrease. The two nerves are antagonistic to each other and are part of the **autonomic nervous system**, which is not under conscious control.

When the blood pressure decreases, or the pH of the blood drops, the receptors send impulses to the cardiac centre. This causes the sympathetic nerve to stimulate the SA node to increase heart rate. Adrenaline from the adrenal gland will also stimulate the SA node and cause an increase in heart rate.

An increase in blood pressure or pH will cause the parasympathetic nerve to be stimulated and the heart rate will fall.

✔ Summary

Control of heart rate

① The diagram shows the mechanism for the control of heart rate.

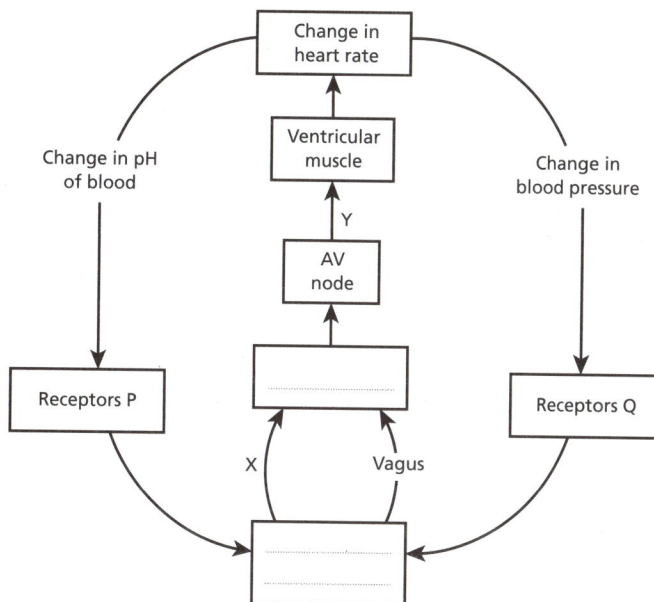

a) Complete the diagram by writing the **two** missing structures in the boxes. [2]

b) Identify the type of receptors shown at P and Q. [2]

P = ...   Q = ...

c) Name the nerve shown at X. .... [1]

d) Describe the pathway that impulses take at Y. [3]

....

....

② The diagram shows the spread of electrical excitation from the sinoatrial node through the heart. The times on the diagram indicate the delay on reaching these structures from the sinoatrial node.

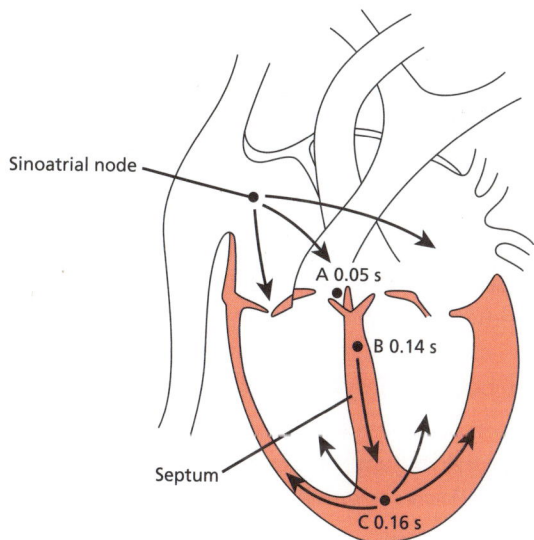

a) Identify structure A. [1]

....

b) The distance from B to C is 0.05 m.

Calculate the speed of conduction of the excitation. [2]

....

c) The speed of the impulse from the sinoatrial node to A is 0.5 ms^{-1}.

Explain the difference between this speed and the speed calculated in part b). [2]

....
....
....

d) Explain why it is important that the excitation moves to the base of the heart before spreading up over the ventricles. [2]

....
....

Nerve impulses

Key words

Key questions

What generates the resting potential?

Why do Na⁺ enter the neurone when channels open?

Why can ions only enter the axon at the nodes of Ranvier?

Action potentials

Like receptors, neurones have a resting potential across their cell membrane, with the inside of the neurone negative compared to the outside. This is maintained by the membrane being impermeable to sodium ions, and pumps in the membrane pumping out three Na^+ in exchange for two K^+ entering. A stimulus causes the potential to become less negative or **depolarise**. If this goes above the threshold, a sequence of events occurs called an **action potential**:

1. A stimulus causes Na^+ channels to open and Na^+ enter the neurone down their electrochemical gradient, causing the membrane to depolarise.
2. If the threshold is reached, other Na^+ channels open and the membrane potential increases and becomes positive.
3. This causes the Na^+ channels to close and K^+ channels open, allowing K^+ to leave the cell down their electrochemical gradient and the membrane repolarises.
4. This process goes beyond −70 mV (**hyperpolarisation**).
5. Finally, the K^+ channels close and the resting potential is restored.

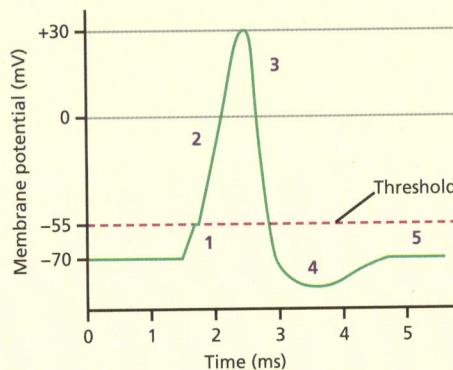

These changes in potential open Na^+ channels further along the membrane, causing the action potential to move along the neurone. If an action potential is generated, it is always the same size. This is the **all or nothing principle**. This means the intensity of a stimulus is not coded by the size of an action potential but by its frequency.

After an action potential has been generated, there is a delay before the Na^+ channels can open again. This is called the **refractory period**.

Myelinated conduction

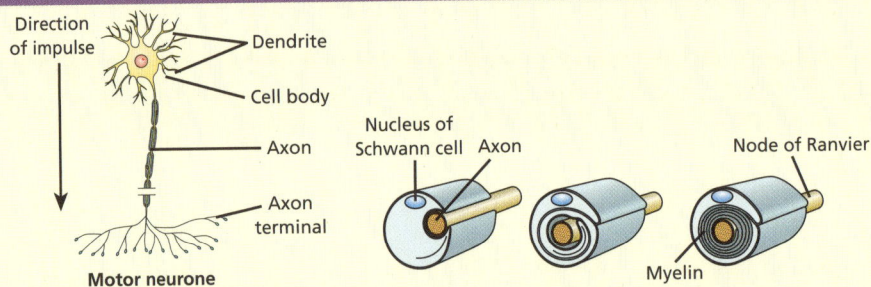

Motor neurones that transmit action potentials from the central nervous system to effectors all possess a myelin sheath and so are said to be **myelinated**.

- The myelin sheath is made from **Schwann cells** that wrap themselves around the axon.
- This produces many layers, mainly composed of phospholipids, from the Schwann cell membrane.
- Gaps called **nodes of Ranvier** are between Schwann cells, where the axon is exposed.

The myelination means that ions can only pass through the axon membrane at the nodes. This causes the action potential to 'jump' from node to node. This is called **saltatory conduction** and it results in faster speeds for the nerve impulse.

Other factors that can affect the speed of conduction are axon diameter (wider axons conduct faster) and temperature (low temperatures reduce speed of ion movements).

Summary

Spec. ref. 3.6.2.1

Nerve impulses

1 The three diagrams, X, Y and Z, show the cell membrane of a neurone at three different times.

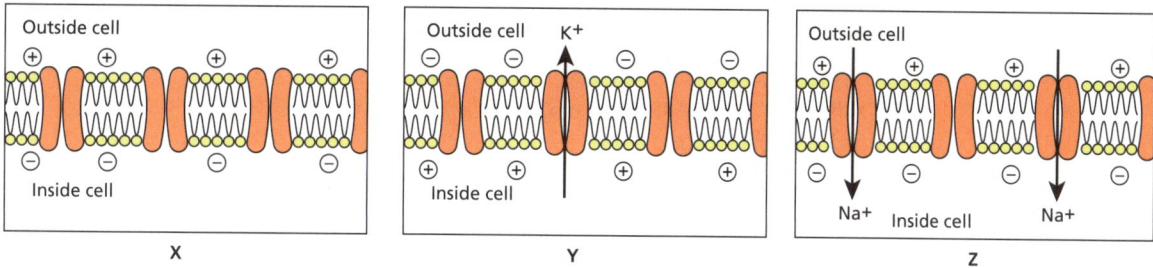

X

Y

Z

Give the letter of the diagram that matches each of these processes occurring in the neurone. [2]

Maintenance of a resting potential = ... Depolarisation = ... Repolarisation = ...

2 The axon of a neurone was subject to six increasing intensities of stimuli at points 1–6. The resulting membrane potential of the axon is shown on the graph.

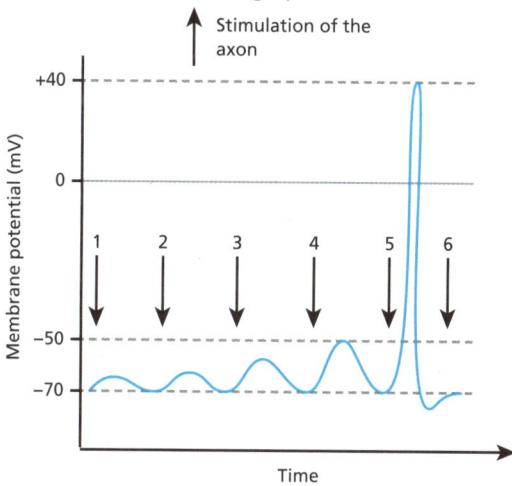

a) Give the resting potential and the threshold potential of the axon. [2]

Resting potential = ...

Threshold potential = ...

b) No action potential was generated after stimuli 1–4. Give the reason why. [2]

....

....

....

c) Explain what produced the increase in membrane potential to +40 mV after stimulus 5. [2]

....

....

d) No action potential was generated by stimulus 6, even though it had a larger intensity than stimulus 5. Explain why. [2]

....

....

3 The graph shows the relationship between axon diameter and conduction velocity in myelinated and unmyelinated neurones.

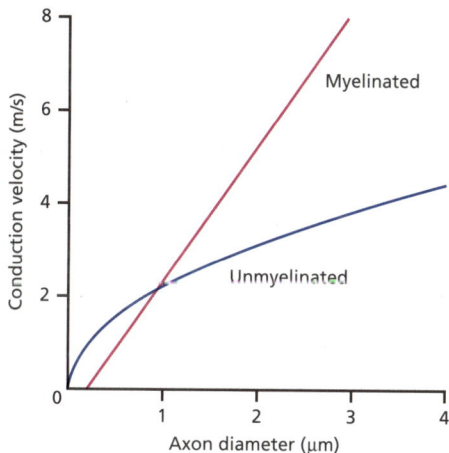

a) Describe the relationship between axon diameter and conduction velocity in each type of neurone. [4]

Myelinated neurones:

....

....

....

Unmyelinated neurones:

....

....

....

b) Explain why myelinated neurones give a faster velocity than unmyelinated neurones for most diameters. [3]

....

....

....

Synaptic transmission

Key words

Key questions

Why is synaptic transmission described as cell signalling?

What would happen if the neurotransmitter was not broken down?

What happens in the brain when a student learns a new biological fact?

Synapses and neuromuscular junctions

Between the end of a neurone and the dendrites of the next neurone, there is a gap called a **synapse**. Synapses use chemicals called **neurotransmitters** to transmit information from one neurone to the next.

In **cholinergic** synapses, the neurotransmitter is **acetylcholine**. However, in the brain there are various other neurotransmitters. When an action potential reaches the **synaptic knob**, various processes take place:

1. The action potential stimulates the opening of Ca^{2+} channels in the synaptic knob membrane.
2. Ca^{2+} enter, causing **synaptic vesicles** to move towards the **presynaptic membrane**.
3. The vesicles fuse with the membrane, releasing the neurotransmitter into the **synaptic cleft**.
4. The neurotransmitter molecules diffuse across the cleft and bind with receptors on the **postsynaptic membrane**.
5. This causes Na^+ channels to open in the membrane and Na^+ enters, depolarising the membrane. If the depolarisation is large enough, an action potential is generated in the postsynaptic neurone.

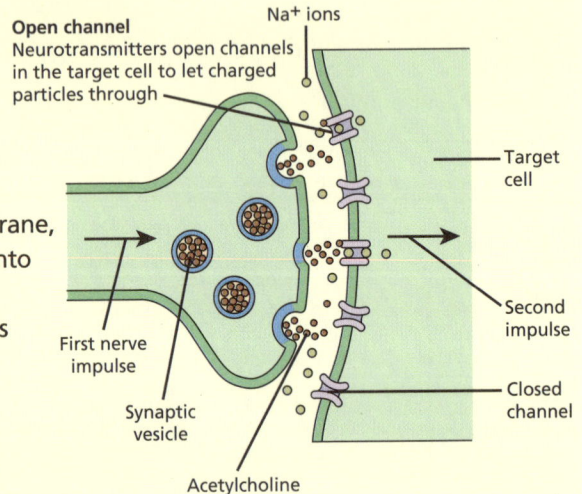

Open channel
Neurotransmitters open channels in the target cell to let charged particles through

Na$^+$ ions

Target cell

Second impulse

Closed channel

First nerve impulse

Synaptic vesicle

Acetylcholine

After the action potential has been generated, acetylcholine is broken down by the enzyme acetylcholinesterase. This ensures that the signal does not continue indefinitely.

Where a motor neurone ends on a muscle fibre, there is a structure like a synapse. This is called a **neuromuscular junction** (NMJ). The mechanism for transmitting an impulse across a synapse and a NMJ is similar, however there are a few differences:

- NMJ are between motor neurones and muscle fibres, but synapses are between two neurones.
- At NMJs, acetylcholine is always the neurotransmitter, but synapses can use different chemicals.
- The postsynaptic membrane is more folded at NMJs.

Processing at synapses

Synapses allow information in the nervous system to be processed in various ways:

- Several neurones may synapse onto one neurone and neurotransmitters from all may be needed to stimulate an action potential – **spacial summation**.
- Several impulses may need to reach a single synaptic knob for it to be able to stimulate an action potential – **temporal summation**.
- There are **inhibitory synapses** in the brain that suppress nerve activity and prevent over-stimulation.

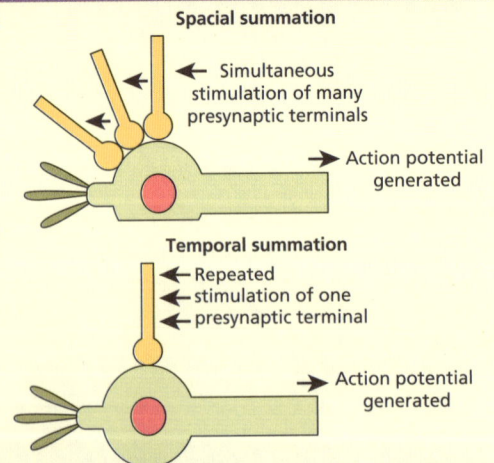

Spacial summation

Simultaneous stimulation of many presynaptic terminals

Action potential generated

Temporal summation

Repeated stimulation of one presynaptic terminal

Action potential generated

Summary

Synaptic transmission

1 The statements are steps in the passage of an impulse across a synapse. They are not in the correct order.

A. Neurotransmitter binds with receptor site

B. Vesicle binds with presynaptic membrane

C. Ca^{2+} enter presynaptic neurone

D. Postsynaptic membrane depolarises

E. Na^+ channels open

F. Neurotransmitter diffuses

Write the letters A–F in the boxes to show the correct order of these steps. [5]

2 The diagram shows a synapse in the brain. The neurotransmitter is dopamine. Dopamine is responsible for activating neurones involved in pleasure pathways. Cocaine is an illegal drug that can result in feelings of pleasure.

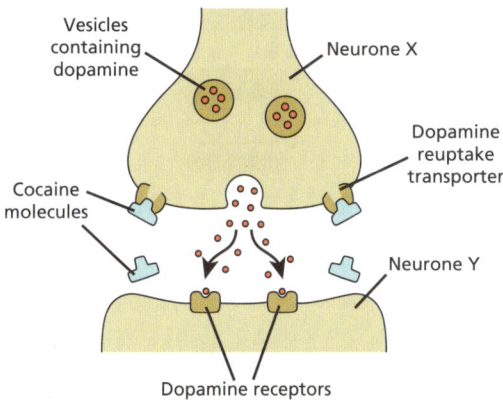

Vesicles containing dopamine

Neurone X

Dopamine reuptake transporter

Cocaine molecules

Neurone Y

Dopamine receptors

a) Use the diagram to explain why nerve impulses can only pass from neurone X to neurone Y, and not in the reverse direction. [2]

b) After stimulating neurone Y, dopamine is taken back into neurone X.

Use the diagram to explain how cocaine can produce feelings of pleasure. [3]

3 The diagram shows the structure of acetylcholine and aldicarb. Aldicarb is used as an insecticide.

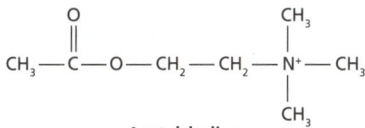

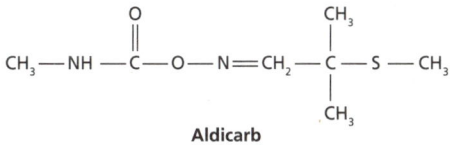

$$CH_3 - C(=O) - O - CH_2 - CH_2 - N^+(CH_3)_2 - CH_3$$

Acetylcholine

$$CH_3 - NH - C(=O) - O - N=CH_2 - C(CH_3)_2 - S - CH_3$$

Aldicarb

a) State the role of acetylcholine at synapses. [1]

b) Aldicarb acts as an inhibitor of acetylcholinesterase. Use the diagram to explain how. [2]

c) Suggest how aldicarb may cause the death of insects. [3]

4 The diagram shows two presynaptic neurones, **A** and **B**, synapsing with one postsynaptic neurone. The presynaptic neurones contain the neurotransmitters acetylcholine or GABA.

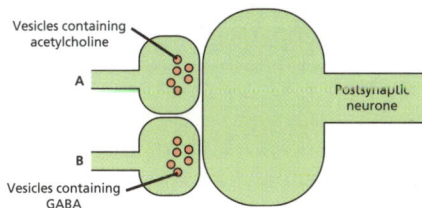

Vesicles containing acetylcholine

A

Postsynaptic neurone

B

Vesicles containing GABA

a) Several action potentials in A arriving at the synapse, one after another, are needed to trigger an action potential in the postsynaptic neurone.

Name this type of interaction. [1]

b) The neurotransmitter GABA causes Cl^- to enter the postsynaptic neurone.

Explain why neurone B acts as an inhibitory synapse. [3]

Muscle contraction

Key words

Key questions

Why does the A band look darker than the I band?

Why does Ca²⁺ need to be pumped back into the vesicles after contraction?

Why do slow muscle fibres have a richer blood supply than slow fibres?

Structure of skeletal muscle

Skeletal muscles act in antagonistic pairs to move the skeleton.

- Skeletal muscle contains many muscle fibres, connective tissue and blood vessels.
- Each muscle fibre contains a number of smaller fibres called **myofibrils**.
- They are made from repeating units called **sarcomeres**, which contain two main proteins called **actin** and **myosin**.
- These proteins are arranged in an overlapping pattern, which produces characteristic striations in the muscle.
- The myosin molecule has heads that can attach to binding sites on the actin fibres, forming a cross-bridge.
- A third protein called tropomyosin sits on the actin molecule and can block the binding sites.
- Throughout the cytoplasm of the fibres is a series of tubes called sarcoplasmic reticulum that contain vesicles of Ca^{2+}.

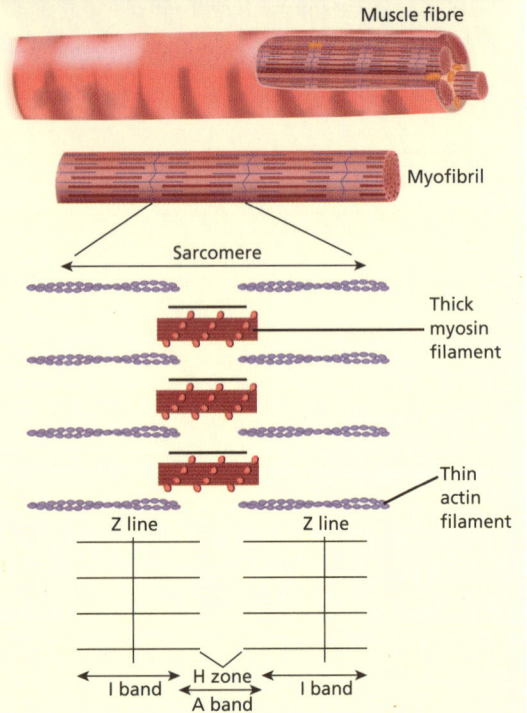

The muscle fibres have structures called motor end plates on them that are the terminals of motor neurones. It is here that the neuromuscular junctions (NMJ) are found.

Muscle contraction

When an impulse reaches the NMJ, the release of acetylcholine stimulates an impulse to spread over the surface of the fibres and deep into them. This causes the release of Ca^{2+} from the sarcoplasmic reticulum. The Ca^{2+} bind with tropomyosin and cause it to expose the binding sites on the actin molecules. This allows these processes to occur:

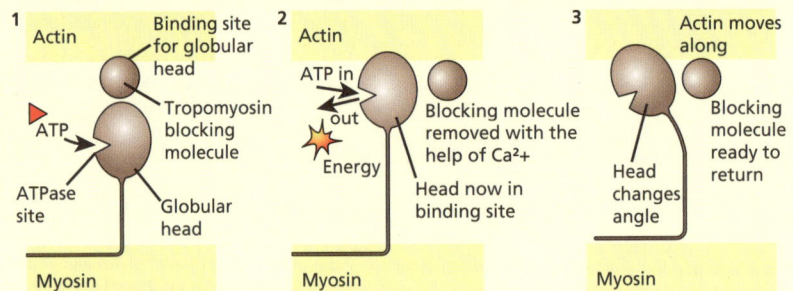

The change in angle of the myosin head causes actin and myosin fibres to slide past each other. This results in the muscle contracting and is called the **sliding filament hypothesis**. The ATP needed for this process can be regenerated from a molecule called **phosphocreatine**, which can phosphorylate ADP.

There are two types of striated muscle fibres. **Fast fibres** contract rapidly, for short periods of time, using anaerobic respiration to produce ATP from glycolysis. **Slow fibres** contract slowly for longer periods of time using aerobic respiration.

Summary

Muscle contraction

1. These statements are steps in the contraction of a skeletal muscle. They are not in the correct order.

A. Muscle membrane is depolarised

B. Actin and myosin cross-bridges form

C. Ca^{2+} released from sarcoplasmic reticulum

D. Tropomyosin moves away from actin

E. Myosin head changes angle

F. Acetylcholine binds with muscle membrane

Write the letters A–F in the boxes to show the correct order of these steps. [5]

2. Which of these statements are true about fast and slow muscle fibres? Tick **one** box. [1]

   1. Fast muscle fibres need large numbers of mitochondria in their cytoplasm.

   2. Muscles that are responsible for maintaining posture have many slow fibres.

   3. Slow fibres have a deeper red colour than fast fibres as they need a richer blood supply.

   1, 2 and 3 ☐   1 and 2 ☐   1 and 3 ☐   2 and 3 ☐

3. The diagram shows the region where a motor neurone reaches a skeletal muscle fibre.

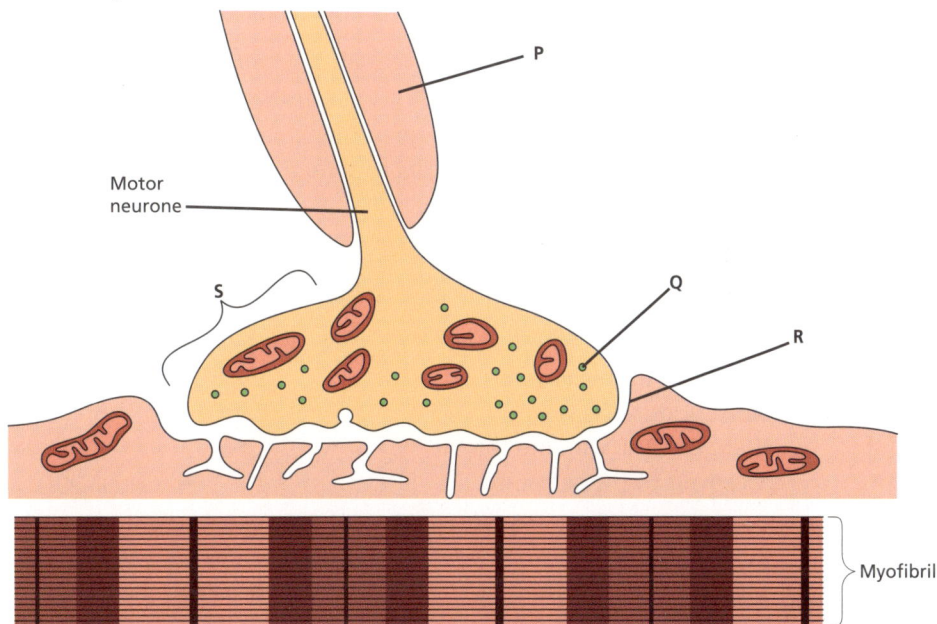

a) Identify the structures labelled P, Q, R and S. [4]

P = ...   Q = ...

R = ...   S = ...

b) On the diagram, label the position of:

   • one sarcomere

   • an A band and an I band

   • an H zone. [4]

c) When the muscle contracts, changes occur in the myofibril.

Complete this table, stating whether the structures get smaller, get larger or stay the same. [4]

A band	
I band	
H zone	
Sarcomere	

Homeostasis and negative feedback

Key words

Key questions

What effect can an increase in water potential have on cells?

What is a set point?

Why is positive feedback not involved in homeostasis?

The importance of homeostasis

Homeostasis is the ability to maintain a constant internal environment despite changes in the external environment. Some organisms, such as mammals, control their internal environment very closely. This requires energy to achieve but it can give them an advantage over organisms that have less control.

Examples of homeostasis in mammals include:

- **Temperature regulation** (thermoregulation) – internal body temperature is maintained within close limits to ensure that metabolic reactions take place under optimal conditions. Any variations in temperature will affect enzyme action.
- **pH regulation** – if the pH of the blood changes then various proteins, such as enzymes, can be structurally altered or denatured.
- **Regulation of blood glucose** – levels of glucose must be maintained within a narrow range. Too little glucose in the blood causes respiration to slow, whereas too much glucose can lead to changes in the water potential of the blood, dehydration and circulatory collapse.
- **Regulation of water levels** (osmoregulation) – if the water potential of the blood changes, this can have osmotic effects on cells.

Negative and positive feedback

Homeostasis requires communication between the different cells in the body. Receptor cells detect any change in the variable from the **set point** or normal level. Any change causes the receptors to send a signal to a control centre. A signal is then sent to effectors, which carry out the required response. The process used is called **negative feedback**. This is defined as: any change in a variable from the set point brings about a change in the variable that returns it to the set point.

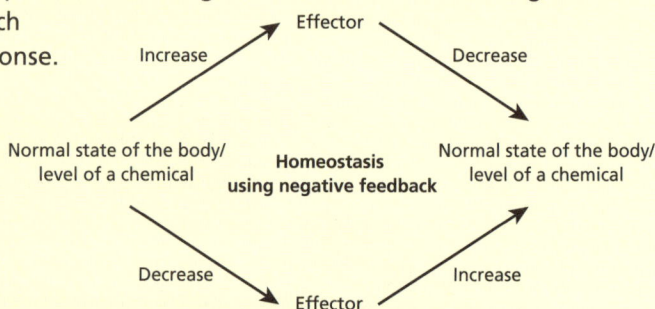

Negative feedback can work with only one control mechanism, a little like a thermostat in a house turning the heat off when it is too hot and on when it is too cold. But two different control mechanisms provide much finer control and smaller fluctuations from the normal value, like the thermostat turning the heat off if it is too hot but also opening windows.

Positive feedback is not involved in homeostasis, as it does not return variables to their normal level. It involves detecting a variation from the set point, which then triggers a response that moves the variable further away from the set point. An example is seen in the production of an action potential (see page 98).

A stimulus causes Na⁺ channels to open, which causes the membrane potential to depolarise. This acts on more Na⁺ channels, opening them, and so further depolarising the membrane.

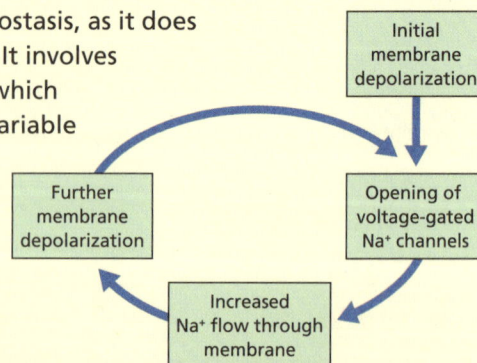

✔ Summary

Spec. ref. 3.6.4.1

Homeostasis and negative feedback

1 The graph shows how a variable in the body changes during a negative feedback process.

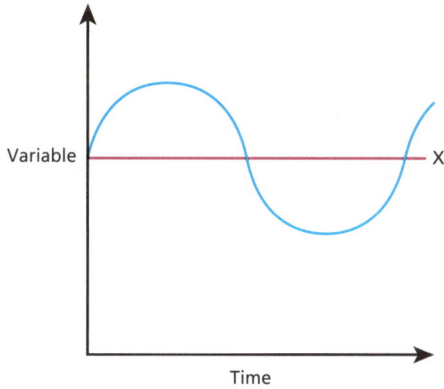

a) Give the name used to describe the value X on the graph. [1]

....

b) The process shown in the graph involves one control mechanism.

Draw a line on the graph to show the variable if two control mechanisms were involved. [1]

2 Thyroxine is a hormone that is released from the thyroid gland in mammals. The diagram shows how the release of thyroxine is controlled.

Use the diagram to explain the type of feedback mechanism that is used to control thyroxine levels in the body. [4]

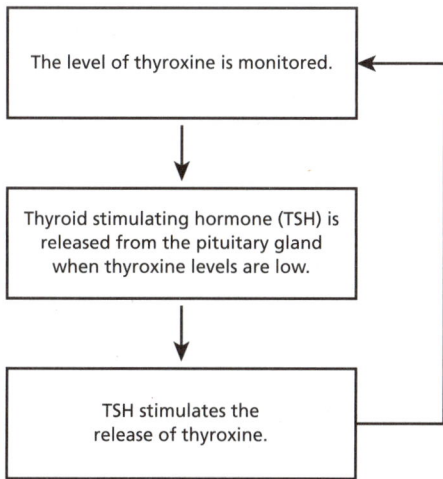

The level of thyroxine is monitored.

↓

Thyroid stimulating hormone (TSH) is released from the pituitary gland when thyroxine levels are low.

↓

TSH stimulates the release of thyroxine.

....

....

....

....

....

....

....

....

....

....

3 The diagram shows the process that regulates childbirth.

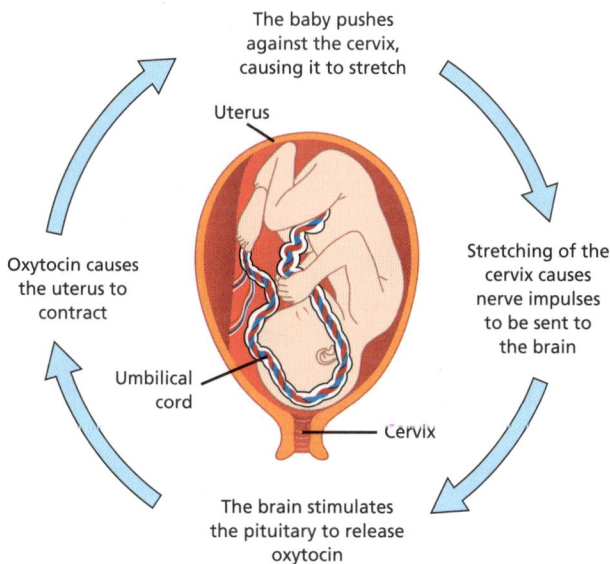

The baby pushes against the cervix, causing it to stretch

Uterus

Oxytocin causes the uterus to contract

Umbilical cord

Cervix

Stretching of the cervix causes nerve impulses to be sent to the brain

The brain stimulates the pituitary to release oxytocin

a) For this process, identify the receptors, coordinator and effector. [3]

Receptors ....

Coordinator ....

Effector ....

b) Use the information in the diagram to explain what type of feedback process is involved. [4]

....

....

....

....

Control of blood glucose

Key words

Key questions

Which two hormones have similar functions?

Where do glycerol and amino acids come from for gluconeogenesis?

Why would exercise decrease blood glucose levels?

Control mechanisms

After a meal, glucose is absorbed, causing the blood glucose concentration to rise. After fasting or exercise, there will be a fall in the level. Much of the control of blood glucose is carried out by the liver using three processes:

- **Glycogenesis**, the formation of the polysaccharide glycogen from glucose.
- **Glycogenolysis**, the hydrolysis of glycogen to produce glucose.
- **Gluconeogenesis**, the production of glucose from non-carbohydrate molecules such as glycerol or amino acids.

Three main hormones are used to control the blood glucose level by negative feedback:

Insulin is produced in the beta cells of the islets of Langerhans in the pancreas. Insulin decreases blood glucose levels:

- It increases the uptake of glucose by cells by inserting channel proteins in cell membranes.
- It activates enzymes involved in glycogenesis.

Glucagon is produced by the alpha cells of the islets of Langerhans in the pancreas. It increases blood glucose levels by:

- Activating enzymes involved in glycogenolysis.
- Activating enzymes involved in gluconeogenesis.

Adrenaline is produced in the adrenal gland and increases blood glucose levels by activating enzymes involved in glycogenolysis. The three hormones cannot pass through the cell membrane and so need to use a system to signal to the inside of the cell. This is shown for glucagon in the diagram:

1. Glucagon binds with a receptor.
2. G protein is activated.
3. The G protein triggers adenylate cyclase to make cAMP.
4. cAMP activates protein kinase, which by a cascade of reactions, causes glycogenolysis.

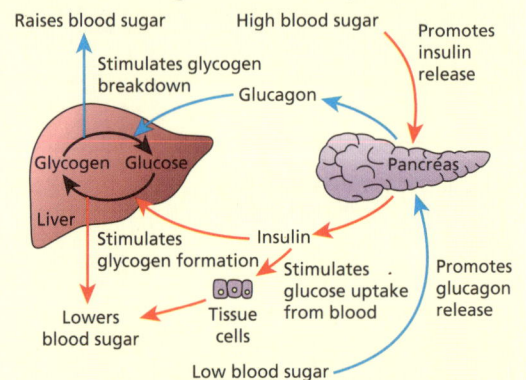

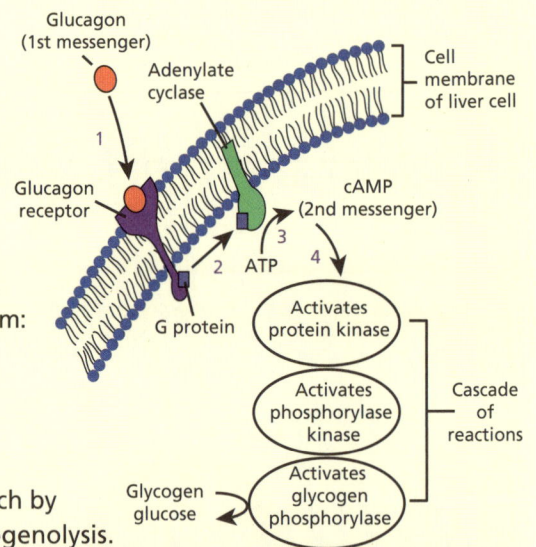

Diabetes

Diabetes mellitus is a condition in which blood glucose levels may become too high. This can result in glucose being passed out in the urine, drawing water with it by osmosis. There are two main types of diabetes:

- **Type I diabetes** is caused by the immune system destroying the beta cells in the pancreas. As a result, the body cannot produce enough insulin.
- **Type II diabetes** is more common and is often caused by the body cells becoming less sensitive to insulin. In many cases this is linked to obesity.

Type I is treated by monitoring blood glucose and injecting insulin. Type II is treated by controlling diet to reduce sugar, exercise to reduce blood glucose levels, and weight loss.

Summary

Control of blood glucose

(1) Complete the paragraph to explain the role of insulin in the control of blood glucose levels. [5]

When the concentration of blood glucose rises above a set point, ... cells in the pancreas secrete

insulin. Insulin binds with .... on the surface of body cells. This causes the incorporation of more

.... into the cell membrane. Liver cells convert glucose to

.... in a process called .... The concentration of blood glucose therefore

returns to normal.

(2) The normal range for blood glucose level is 3.9–5.5 mmol dm^{-3}. In a glucose tolerance test, a patient is given glucose to drink and their blood glucose and blood insulin levels are measured. The graph shows the results of a test on a patient.

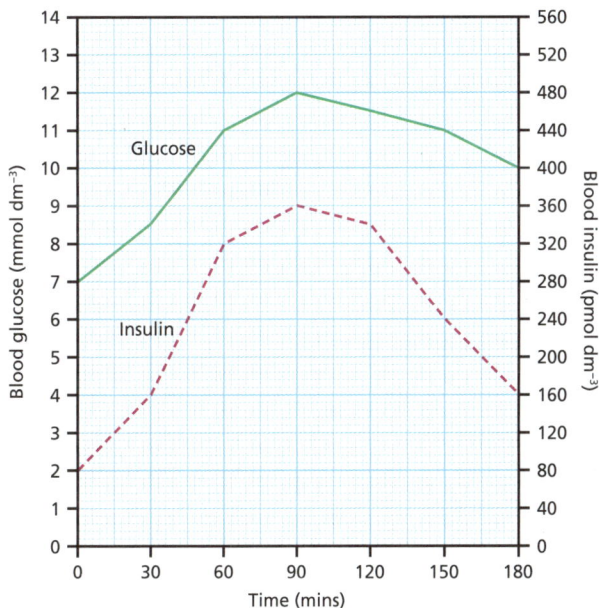

a) State why it is important that the person has not eaten for six hours before the test. [1]

....

b) By looking at the blood glucose level at time 0, the doctor concludes that the patient has diabetes. Give a reason why. [2]

....

....

....

....

c) The doctor decides that the patient has type II diabetes and not type I.

What evidence is there for this from the graph? [2]

....

....

....

(3) The graph shows the activity of two enzymes in the liver of a person taken after they have consumed a solution of glucose.

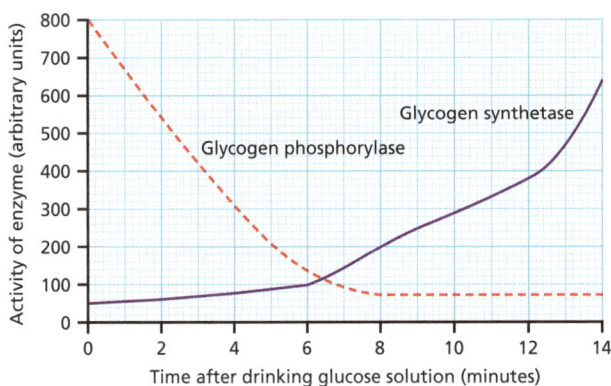

a) Calculate the percentage increase in the activity of glycogen synthetase from 0 to 14 minutes. [2]

....

b) Name the hormone responsible for this change.

.... [1]

c) Describe the action of glycogen synthetase. [2]

....

....

d) Glycogen phosphorylase is activated by the actions of glucagon.

Explain how the release of glucagon leads to the activation of glycogen phosphorylase. [5]

....

....

....

....

Key words

Key questions

Why do blood cells stay in the blood?

The nephron

The kidneys are involved in excretion and in the control of the water potential of the blood. They contain thousands of microscopic tubules called **nephrons**.

- Blood from the renal artery enters a small knot of blood capillaries called the **glomerulus**.
- Due to the pressure of the blood, it is filtered and small molecules pass out; this is **ultrafiltration**.
- The **glomerular filtrate** then collects in the **Bowman's capsule**.
- In the **proximal convoluted tubule**, all the glucose and amino acids are selectively reabsorbed back into the blood.

A nephron

Renal corpuscle — Proximal convoluted tubule — Distal convoluted tubule — Efferent arterioles — Glomerulus — Afferent arterioles — Bowman's capsule — Renal cortex — Renal medulla — Descending limb (permeable to water) — Loop of Henle — Collecting duct — Vasa recta — Ascending limb (permeable to salts)

What is the function of the loop of Henle?

The filtrate then flows into the **loop of Henle**, which has the function of creating a concentration gradient in the tissue of the kidney.

- In the ascending limb of the loop, Na^+ is pumped out of the tubule and diffuses into the bloodstream.
- This decreases the water potential of the blood, so water is drawn out of the descending limb.
- The effect of this is to produce a concentration gradient in the kidney tissue, with the areas near the base of the loop having a very low water potential.
- This is called a **counter-current multiplier** system because the blood and fluid flow in opposite directions, increasing the size of the concentration gradient produced.

Counter-current flow

Descending Ascending
Ascending   Descending
Na^+ H_2O Na^+ Na^+
H_2O Na^+ H_2O

Osmoregulation

Up to the end of the loop of Henle, all the processes occur irrespective of the water potential of the blood. It is the distal convoluted tubule and collecting duct that regulate the concentration of the blood. This is called **osmoregulation**.

Why is ADH called antidiuretic?

The water potential of the blood is detected by **osmoreceptors** in the **hypothalamus**. If the water potential is too low, the hypothalamus sends nerve impulses to the posterior pituitary gland to increase the release of **antidiuretic hormone** (ADH).

ADH increases the permeability of the walls of the collecting ducts and distal convoluted tubule by opening **aquaporin** channels that let water pass down an osmotic gradient into the blood. A smaller volume of more concentrated urine is then produced.

When the blood is too dilute, the hypothalamus triggers a reduction in ADH release.

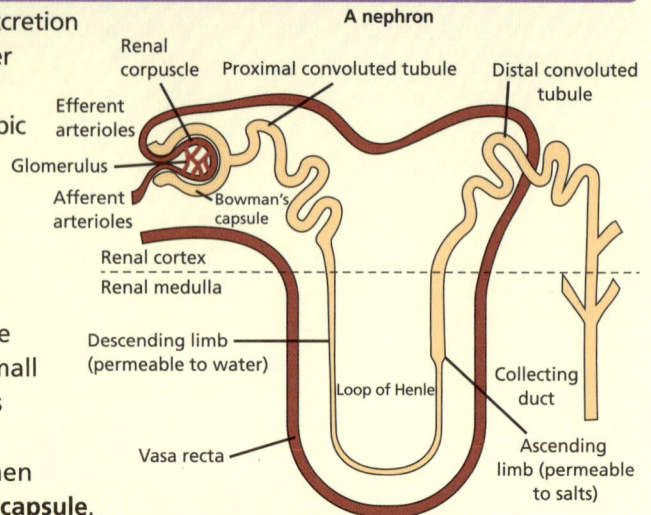

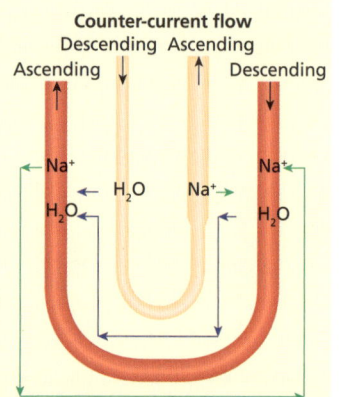

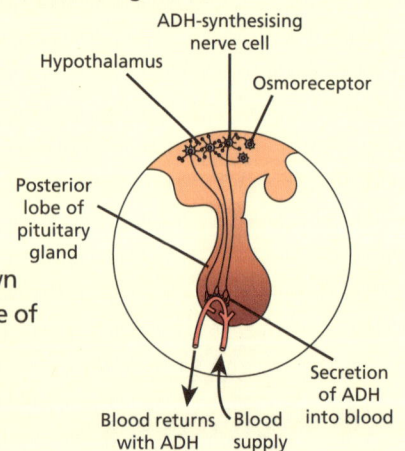

ADH-synthesising nerve cell — Hypothalamus — Osmoreceptor — Posterior lobe of pituitary gland — Secretion of ADH into blood — Blood returns with ADH — Blood supply

Summary

Control of blood water potential

1. Where is ADH produced? Tick **one** box. [1]

In the distal convoluted tubule ☐   In the hypothalamus ☐

In the pituitary gland ☐   In the collecting ducts ☐

2. The table shows the composition of three different fluids found in the kidney.

Component	Concentration (g 100 cm^{-3})		
	Blood plasma entering glomerulus	Glomerular filtrate	Urine in collecting duct
Water	90–93	97–99	96
Proteins	7–9	0.0	0.0
Glucose	0.1	0.1	0.0
Urea	0.03	0.03	2.0

a) Explain the protein content of the glomerular filtrate. [2]

b) Explain the change in water content between the glomerular filtrate and the urine. [2]

c) Urea is not pumped into the urine.

Explain the change in concentration of urea between the glomerular filtrate and the urine. [2]

3. The diagram shows a cell from the lining of the proximal convoluted tubule.

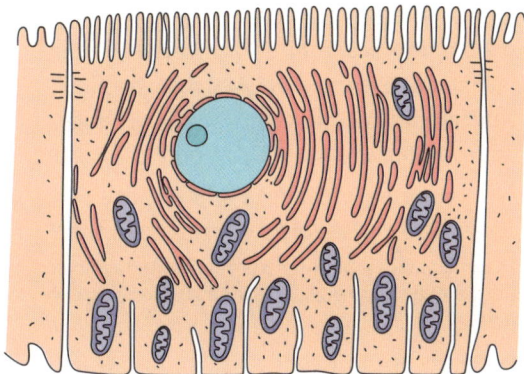

a) Explain **two** features shown in the diagram that make the cell adapted for the selective reabsorption of glucose. [4]

b) As well as reabsorbing glucose and amino acids, water is also reabsorbed back into the blood in the proximal convoluted tubule. Explain why this happens. [2]

4. The diagram shows a human nephron and a gerbil nephron, drawn to the same scale.

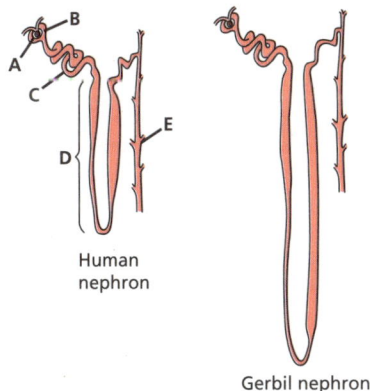

Human nephron

Gerbil nephron

a) Identify the parts labelled A to E on the human nephron. [5]

A = ...   B = ...

C = ...   D = ...

E = ...

b) Gerbils live in hot dry deserts. Explain how the difference between their nephrons and human nephrons allow gerbils to live in these conditions. [4]

Inheritance with monohybrid crosses

🔑 Key words

❓ Key questions

What is the difference between a gene and an allele?

Genes, alleles and monohybrid inheritance

Diploid individuals have two copies of each chromosome. Therefore, they have two copies of each gene, called **alleles**. If both alleles are the same then the individual is **homozygous** for that gene. If they are different then they are **heterozygous**. The expression of the gene in the individual is called the **phenotype** and the actual alleles that are present is called the **genotype**.

Some alleles are **dominant**, meaning they will always be expressed in the phenotype when present. Others are **recessive** and are only expressed when both alleles are recessive. For some genes, alleles can be **codominant**, meaning both are expressed in the heterozygote.

In monohybrid inheritance, a characteristic is controlled by a single gene. This can be shown using Punnett squares. The diagram shows Punnett squares for cystic fibrosis (CF). F represents the dominant allele and f is the recessive allele. The disorder is caused by the recessive allele.

	F	F
F	FF	FF
F	FF	FF

	F	f
F	FF	Ff
f	Ff	ff

	F	F
F	FF	FF
f	Ff	Ff

Both parents are homozygous dominant and do not carry the CF allele. No offspring will have or carry the condition.

Each parent is heterozygous. They carry the CF allele. The phenotype of the offspring will be 3 without CF: 1 with CF.

One parent is free of the CF allele; one is a carrier. All offspring will be free of CF but 50% will carry the CF allele.

How many phenotypes can there be in a monohybrid, codominant cross?

Codominance can also be shown using Punnett squares. An example is coat colour in some cattle. The allele C^R codes for a red coat colour and C^W codes for a white colour. An individual that has the genotype $C^R C^W$ has an intermediate colour called roan.

Chi-squared tests

The chi-squared (χ^2) test is used to test for a significant difference between what has been observed in an experiment and what was expected. It gives a probability (p) value, which is an indication of whether any difference is significant or due to chance. If it is high (a low χ^2 value) then the obtained result occurred due to random chance alone.

The test can only be used when there are categorical data and is quite often used to analyse the results of genetic crosses. The chi-squared (χ^2) test uses the formula:

$$\chi^2 = \sum \frac{(\text{Obs} - \text{Exp})^2}{\text{Exp}}$$

(where Obs is the observed result, e.g. the offspring counted with a particular phenotype, and Exp is the expected result, e.g. the offspring you expect to have that phenotype.)

What does a high chi-squared value indicate?

To carry out a chi-squared test, these steps are followed:

1. State the null hypothesis, i.e. there is no difference between the observed and expected result.
2. Calculate the expected number of each phenotype (assume they follow a certain ratio).
3. Calculate the value of χ^2
4. Work out degrees of freedom: in a cross, number of phenotype groups minus 1.
5. Look up the critical value at the 5% (0.05) p value in a χ^2 table for the given degrees of freedom.

If the calculated χ^2 value is **lower** than the critical value, the null hypothesis is accepted, so there is no significant difference between the observed and expected results. If it is **higher** than the critical value, there is a significant difference between them.

✔ Summary

Inheritance with monohybrid crosses

1 In snapdragon plants, flower colour is controlled by a single gene with codominant alleles coding for white or red.

If two pink-flowered plants are bred together, what ratio of plants will be produced? Tick **one** box. [1]

All pink-flowered ☐	1 red-flowered : 1 white-flowered : 1 pink-flowered	☐
3 red-flowered : 1 white-flowered ☐	1 red-flowered : 1 white-flowered : 2 pink-flowered	☐

2 In a cross between two pea plants with red flowers, a plant breeder obtained 105 red-flowered plants and 45 white-flowered plants.

a) Complete the Punnett square to show how this result could be obtained. [2]

	Red parent		
Red parent			

b) The plant breeder wanted to tell the difference between heterozygous and homozygous, red-flowered plants. Explain a breeding experiment that would allow them to do this. [3]

....

....

c) The breeder carries out a chi-squared test on the results of their original cross to see if they fit a 3 : 1 ratio.

i) Complete the table by calculating the expected values. [2]

Phenotype	Observed	Expected
Red	105	
White	45	

ii) Use the chi-squared formula on page 110 to calculate χ^2 [3]

....

iii) Use this section from a χ^2 table to explain what the plant breeder can conclude about this cross. [3]

....

....

....

....

....

Degrees of freedom	Probability value			
	0.10	0.05	0.01	0.001
1	2.71	3.84	6.63	10.83
2	4.61	5.99	9.21	13.82
3	6.25	7.81	11.34	16.27

3 The diagram shows a family pedigree showing the incidence of a condition called albinism.

a) Give evidence from the diagram to show whether albinism is caused by a dominant or recessive allele. [2]

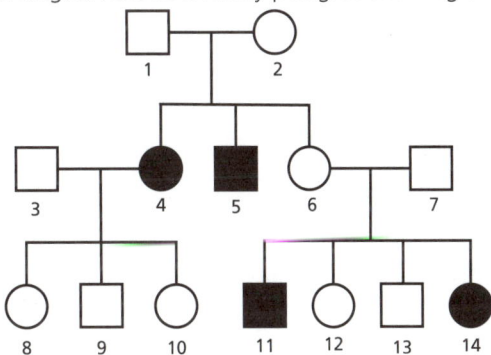

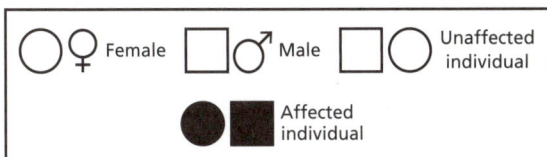

....

b) What percentage of the people in the pedigree are:

i) homozygous recessive? [2]

ii) definitely heterozygous? [2]

c) If individuals 10 and 11 had a child, calculate the chance of it being a boy with albinism. Show how you worked out your answer. [3]

Key:
○♀ Female   □♂ Male   □ ○ Unaffected individual
● ■ Affected individual

Dihybrid ratios and sex linkage

Key words

Key questions

What is a normal dihybrid ratio for double heterozygote crosses?

What are sex chromosomes?

Why does a female have two copies of sex-linked genes?

Dihybrid inheritance

Many genetic crosses involve investigating two different characteristics, controlled by separate genes, carried on different chromosomes. This is called **dihybrid inheritance**.

Each contributing gene has two alleles, which are represented in the same way as for monohybrid crosses, i.e. with a capital letter for the dominant allele and a lower-case letter for the recessive allele. Once again, Punnett squares are used to predict the possible offspring.

In this example, peas can be yellow (Y, dominant) or green (y, recessive), and round (R, dominant) or wrinkled (r, recessive).

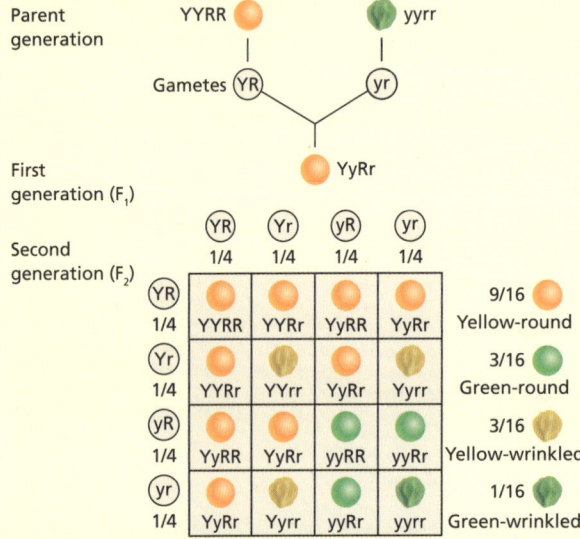

Parent generation: YYRR, yyrr
Gametes: YR, yr
First generation (F_1): YyRr

Second generation (F_2):

	YR 1/4	Yr 1/4	yR 1/4	yr 1/4
YR 1/4	YYRR	YYRr	YyRR	YyRr
Yr 1/4	YYRr	YYrr	YyRr	Yyrr
yR 1/4	YyRR	YyRr	yyRR	yyRr
yr 1/4	YyRr	Yyrr	yyRr	yyrr

9/16 Yellow-round
3/16 Green-round
3/16 Yellow-wrinkled
1/16 Green-wrinkled

Due to independent segregation, the double heterozygous individuals in the F_1 generation can produce four types of gametes, containing four possible combinations of alleles.

So a cross between two double heterozygotes produces a typical dihybrid ratio of 9 : 3 : 3 : 1

Sex linkage

Some genes have a locus on one of the sex chromosomes (usually the X chromosome, as this is much larger than the Y chromosome).

This means that the expression of the gene may depend on the sex of the individual. This is termed **sex linkage**.

In humans, as only males have a Y chromosome, any alleles on the Y chromosome can only be passed to male offspring.

If an allele is present on the X chromosome, either females or males can inherit the trait. However, females will have two alleles for this gene but males will only have one.

Female

	X^T	X^t
Male X^T	$X^T X^T$	$X^T X^t$
Y	$X^T Y$	$X^t Y$

If a female is a carrier of a disorder, t, her sons have a 50% chance of having the condition. Her daughters will not have the condition, but may carry it.

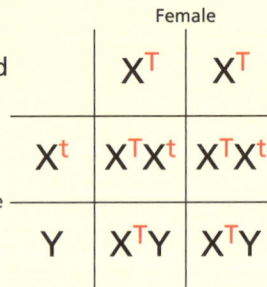

Female

	X^T	X^T
Male X^t	$X^T X^t$	$X^T X^t$
Y	$X^T Y$	$X^T Y$

If a male has an X-linked recessive allele and his partner is homozygous dominant, then all daughters will be carriers. No sons will carry or have the condition as the father does not contribute the X chromosome.

Examples of sex linkage involving the X chromosome include colour blindness, some forms of haemophilia and Duchenne muscular dystrophy.

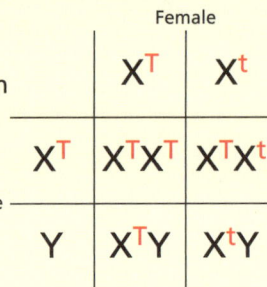

Summary

Dihybrid ratios and sex linkage

(1) In guinea pigs, the allele for black coat (B) is dominant to the allele for white coat (b). The allele for a rough coat (R) is dominant to the allele for a smooth coat (r). The two genes controlling coat texture and colour are on different chromosomes, but not on the sex chromosomes.

a) Give the possible genotypes of each of these phenotypes of guinea pig. [2]

Black with rough coat = ...

White with rough coat = ...

b) A black guinea pig with a rough coat was crossed with a white guinea pig with a smooth coat. All four possible combinations of offspring were produced.

i) Give the genotypes of the two parent guinea pigs. [2]

Black with rough coat = ... White with smooth coat = ...

ii) Complete this genetic diagram to show the expected phenotypic ratio of offspring for this cross.
Use R = rough, r = smooth, B = black and b = white. [3]

Parental genotypes =

Gametes =

Expected ratio of phenotypes = ...

(2) The diagram shows a pedigree for the disorder haemophilia. This disorder is caused by a recessive allele (X^h), which is found on the X chromosome.

□ = male   ○ = female   ■ = male with haemophilia

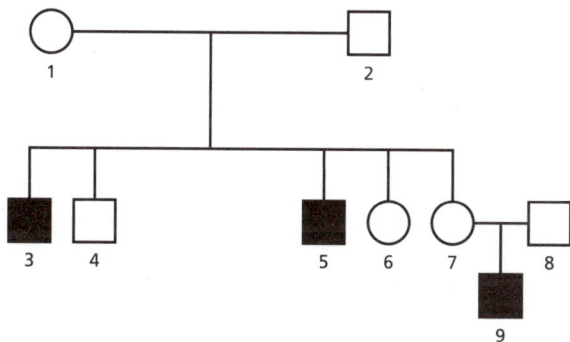

a) Give the genotype of these individuals. The first one has been done for you. [5]

$1 = X^h X^H$ $2 = $ ....

$3 = $ .... $5 = $ ....

$7 = $ .... $8 = $ ....

b) Give the probability of person 6 being a carrier for haemophilia. [1]

....

c) If individuals 7 and 8 had another child, calculate the chance of it having haemophilia. Show how you worked out your answer. [3]

....

d) Explain why it is **not** possible to be a male and a carrier for haemophilia. [3]

....

....

....

Deviations from usual dihybrid ratios

Key words

Key questions

Why does linkage prevent independent segregation?

Autosomal linkage

The **autosomes** are all the chromosomes, apart from the sex chromosomes. If two genes have loci on the same chromosome then their alleles will be inherited together. This is called **autosomal linkage**. As both the genes are on the same chromosome, their alleles cannot show independent segregation. This limits the variety of gametes that can be produced.

In this example, a variety of plant can have red petals (R) or blue petals (r). They can have long stems (L) or short stems (l).

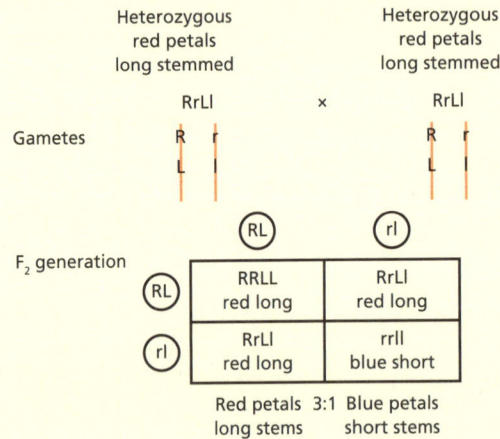

Heterozygous red petals long stemmed

RrLl × RrLl

Gametes: R r / L l

F$_2$ generation

	RL	rl
RL	RRLL red long	RrLl red long
rl	RrLl red long	rrll blue short

Red petals long stems 3:1 Blue petals short stems

In this cross between two double heterozygotes, a 9 : 3 : 3 : 1 ratio would be expected.

However, because the genes are linked, each parent only produces two types of gamete by meiosis. Therefore, a 3 : 1 ratio is produced, without any red-short or blue-long plants.

In reality, a small number of the red-short and blue-long plants would be produced. Typical numbers could be: 603 red-long : 198 blue-short : 10 red-short : 8 blue-long

This happens due to crossing over in meiosis, which produces a small number of the missing gametes and plants. These plants are called **recombinants**.

What are recombinants?

Epistasis

Epistasis is where the expression of one gene depends on another gene. For example, in sweet pea plants two genes are responsible for the production of a purple pigment.

Gene A controls the first step in the reaction to produce a white pigment. Gene B controls the second step in the reaction. If gene A has two recessive alleles, the flowers will be white even if dominant B alleles are present.

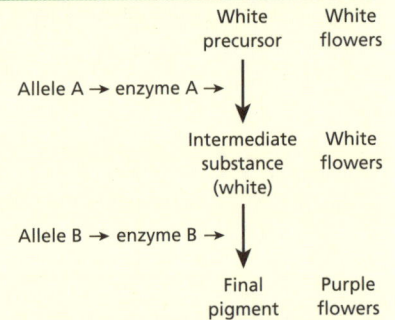

White precursor → White flowers

Allele A → enzyme A →

Intermediate substance (white) → White flowers

Allele B → enzyme B →

Final pigment → Purple flowers

Why can an individual only possess two alleles for each gene?

Multiple alleles

In the examples so far, there have been only two possible alleles for each condition. In many genes there may be **multiple alleles**. But, each individual can only possess two of these alleles.

An example is ABO blood groups in humans. The gene for blood group has three alleles, I^A for A group, I^B for B group and I^O for O group. I^A and I^B are codominant but both are dominant over I^O. The possible genotypes and phenotypes are shown in the table.

Genotypes	Phenotypes (blood groups)
I^A I^A	A
I^A I^B	AB
I^A I^O	A
I^B I^B	B
I^B I^O	B
I^O I^O	O

Summary

Deviations from usual dihybrid ratios

1. In tomato plants, the allele for tall (T) is dominant over the allele for short (t). The allele for green leaves (G) is dominant over the allele for mottled leaves (g).

   A plant breeder crossed a double heterozygous tomato plant and a short plant with mottled leaves.

   The offspring produced were: 25 tall, green plants and 24 short, mottled plants

   a) The **expected** ratio, if this was a standard dihybrid cross, would be 1 : 1 : 1 : 1

      Give an explanation for the ratio obtained by the plant breeder. [3]

   b) In a second cross, a small number of tall, mottled and short, green plants were produced.

      i) Explain how these plants were produced. [2]

      ii) What name is given to individuals that are produced in this way? [1]

2. In some strains of chicken, feather colour can be white or coloured and is controlled by two unlinked genes. The dominant allele C is needed to develop coloured feathers. The allele c codes for white feathers. The dominant allele W produces white feathers, even if the allele C is present.

   a) State the name for this type of gene interaction. [1]

   b) A white chicken of genotype CcWw is crossed with a chicken of genotype ccww.

      Draw a genetic diagram to show the gametes, the genotypes produced and the expected ratio of phenotypes. [4]

      Parents =

      Gametes =

      Expected ratio of phenotypes =

3. The diagram shows a pedigree for human ABO blood groups.

   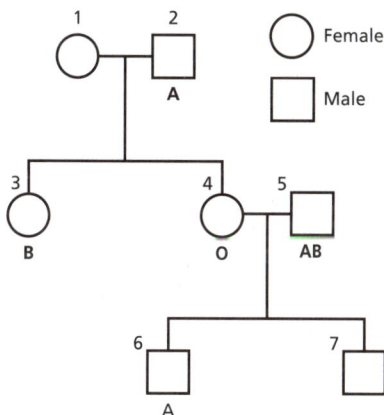

   a) Give the genotype and phenotype of person 1. [2]

      Genotype =

      Phenotype =

   b) Give the possible blood groups for person 7. [1]

   c) If individuals 4 and 5 have another child, calculate the chance that it would be a female, with blood group A. Show how you worked out your answer. [4]

Population genetics

🔑 Key words

Gene pools and the Hardy–Weinberg principle

Genetic crosses can show how genes and their alleles are passed between individuals. Population genetics, however, looks at the genes and alleles of an entire population. Two important terms are used when studying population genetics:

- A **gene pool** is all the alleles of all the genes of all the individuals in a population at any one time
- The **allele frequency** describes the number of times an allele occurs within a particular gene pool of a population.

The term population is important. A population describes all the individuals of the same species living together in a habitat at the same time. As they are members of the same species, they can interbreed with each other to produce fertile offspring (see page 74). Individuals can therefore interchange genes and so share the same gene pool.

Looking at a gene with two alleles, A and a:

- In the whole population, the total allele frequency for the gene has the value 1.0
- If everyone had the genotype AA, then the frequency of the dominant allele (A) would be 1.0 and the frequency of the recessive allele (a) would be 0.0
- If everyone was heterozygous (Aa), the frequency of the dominant allele (A) would be 0.5 and the frequency of the recessive allele (a) would be 0.5

In actual populations, there will usually be a mixture of all three genotypes and the proportions will vary from population to population. It is possible to work out the proportions of genotypes using the **Hardy–Weinberg principle**.

❓ Key questions

What happens if a gene pool becomes very small?

Why would non-random mating negate the Hardy–Weinberg principle?

The Hardy–Weinberg principle says that the allele and genotype frequencies in a population will remain constant from generation to generation. It allows a prediction to be made about the proportion of genotypes in a population in the future. It assumes that:

- no mutations occur
- no natural or artificial selection takes place
- random mating occurs
- no immigration or emigration happens
- there is no genetic drift (see page 120).

(At least one of these factors is usually present in a population, so the principle is often predicting a theoretical number.)

Why does $p + q = 1$ in the Hardy–Weinberg principle?

The Hardy–Weinberg principle is represented by the equation:

$$p^2 + 2pq + q^2 = 1$$

where: p = the frequency of the dominant allele in the population

q = the frequency of the recessive allele in the population

So, for a gene with two possible alleles, $p + q = 1$

p^2 = the frequency of homozygous dominant individuals

q^2 = the frequency of homozygous recessive individuals

$2pq$ = the frequency of heterozygous individuals

A population has been sampled and the frequency of a homozygous recessive genotype (gg) is found to be 36%. What are the frequencies of the G and g alleles and the GG and Gg genotypes?

$p^2 + 2pq + q^2 = 1$

The frequency of gg is 36% (from the question). To calculate the frequency of the g allele, use q^2

$q^2 = 0.36$ $q = \sqrt{0.36}$ So, $q = 0.6$

This means the frequency of the g allele is 60% in the population.

As $q = 0.6$, using $p + q = 1$ means that p must be 0.4, or 40% of the population.

The genotype GG is represented by p^2,

Therefore: $p^2 = 0.4^2$ $p^2 = 0.16$

So, 16% of the population has the GG genotype.

The heterozygous phenotype is given by $2pq$,

so: $2pq = 2(0.4 \times 0.6)$ $2pq = 0.48$

This means that 48% of the population is heterozygous, Gg.

✔ Summary

Population genetics

1. Which organisms share the same gene pool? Tick **one** box to show which statements are correct. [1]

1. All members of the same genus living in the same habitat

2. All individuals of the same species that have existed over time

3. Individuals that reproduce with each other to produce fertile offspring

4. Members of closely related species that mate to produce sterile hybrids

2, 3 and 4 only ☐   2 and 4 only ☐   3 and 4 only ☐   3 only ☐

2. A population of land snails lives in a field. Shell colour is controlled by two alleles. The allele for red shell (R) is dominant and the allele for yellow shell (r) is recessive.

a) A cross between two red-shelled snails gives red-and-yellow-shelled snails.

Give the genotypes of the two parent snails. [1]

b) In a large field, there are 3000 snails and the frequency of the r allele is 0.3

i) Use the Hardy–Weinberg equation to calculate the number of snails that are red and the number that are yellow. [3]

ii) Birds have been introduced into the area. The birds use rocks to break open the shells of the snails so that they can eat them.

Scientists noticed that there were more broken red shells than yellow ones near the rocks. Explain how this may change the allele frequencies of R and r in the field. [2]

3. Thalassaemia is an inherited disorder that affects the body's ability to make haemoglobin. It is caused by a recessive allele. People who are heterozygous for thalassaemia are less likely to get the disease malaria.

The graph shows how the frequency for the thalassaemia allele has changed over the last 5000 years in a population in Africa.

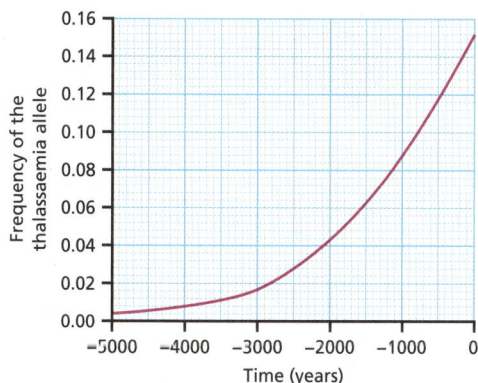

a) Calculate the percentage of people who are carriers for thalassaemia in this population now. Use the Hardy–Weinberg equation and the graph. [3]

b) Explain why the Hardy–Weinberg equation and this graph could **not** be used to calculate the percentage of carriers:

i) in a population in the UK [2]

ii) in the African population in 1000 years' time. [2]

Evolution and speciation

Evolution

Individuals within a population of a species often show a wide range of variation in phenotype. This can be due to genetic or environmental factors (see page 70).

Genetic variation results from mutations that can produce new alleles. Meiosis and the random fertilisation of gametes then combine alleles in different combinations.

Natural selection occurs because of processes such as predation, disease and competition that produce a selection pressure. Those organisms with selective advantages are more likely to produce offspring and pass on their favourable alleles to the next generation. When Charles Darwin developed the theory of natural selection (see page 70), the emphasis was on individuals. Modern population genetics focuses more on the effect on the gene pool. Any differential reproductive success will cause a change in the allele frequencies within a gene pool, which results in evolution.

Which individuals are selected against in disruptive selection?

The effects of stabilising and directional selection are discussed on page 72. However, another type of selection can occur called **disruptive selection**. This occurs when natural selection favours both extremes of variation. Over time, the two extreme variants become more common, whereas the intermediate states become less common or lost altogether.

In terms of evolution, disruptive selection is probably the most important driving force. A classic example has been observed in the peppered moth. The moth can occur in different colours but the most common forms are pale or dark:

- In areas where tree bark is dark, the darker moths are selected for.
- In areas where tree bark is light, the paler moths are selected for.

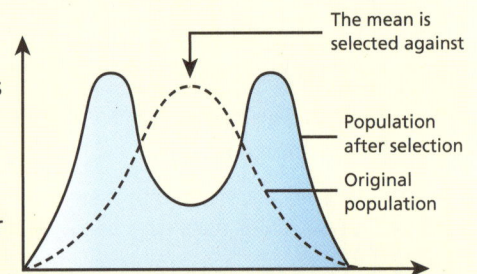

The mean is selected against

Population after selection

Original population

What is the selection pressure acting on peppered moths?

Speciation

Members of the same species can freely interbreed and produce fertile offspring. So if they are members of the same population, there is a gene flow between individuals. But if groups of the population become isolated from each other then, over time, mutations leading to new alleles will arise and spread through the two new populations.

The new alleles may be different in each population and may eventually prevent individuals from each group mating and producing fertile offspring. This means a new species has been formed. This production of a new species from an existing species is called **speciation**.

An **isolating mechanism** is needed to allow this to occur. If the two groups from the original population continue to mate, there will be gene flow and speciation cannot occur.

Geospiza magnirostris

Geospiza conirostris

Geospiza parvula

Certhidea olivacea

Ancestral finch

What will an isolating mechanism prevent?

A classic example of speciation was studied by Darwin on the Galápagos Islands. He found 17 different species of finches on the different islands. He realised that these birds had a common ancestor but were exposed to different selection pressures on different islands. The isolated populations evolved differently and became separate species.

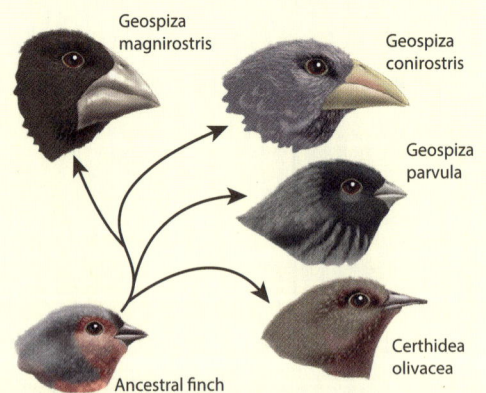

Evolution and speciation

1 What is needed for speciation to occur? Tick **one** box to show which statements are correct. [1]

1. A factor causing reproductive isolation

2. Stable environmental conditions

3. Genetic variation

4. Free gene flow

1, 2, 3 and 4 ☐   1, 3 and 4 only ☐   1 and 3 only ☐   2 and 4 only ☐

2 Scientists investigated selection in spadefoot toads in ponds in Arizona. Spadefoot toads reproduce and the eggs develop into tadpoles.

Scientists have observed that the tadpoles can feed on different foods and have different adaptations:

- Those with larger mouths and jaw muscles tend to be carnivores.
- Those with smaller mouths and jaw muscles feed mainly on plants.

The scientists gave each tadpole a number from 1 to 5, to indicate their type of feeding, with 1 being more herbivore-like and 5 being more carnivore-like. They also measured the mass of each tadpole.

a) The scientists used the mass of the tadpoles as an indication of their 'fitness', with a greater mass indicating a better fitness.

Suggest what the scientists meant by 'fitness'. [2]

b) The graphs show the scientists' results from two ponds, A and B.

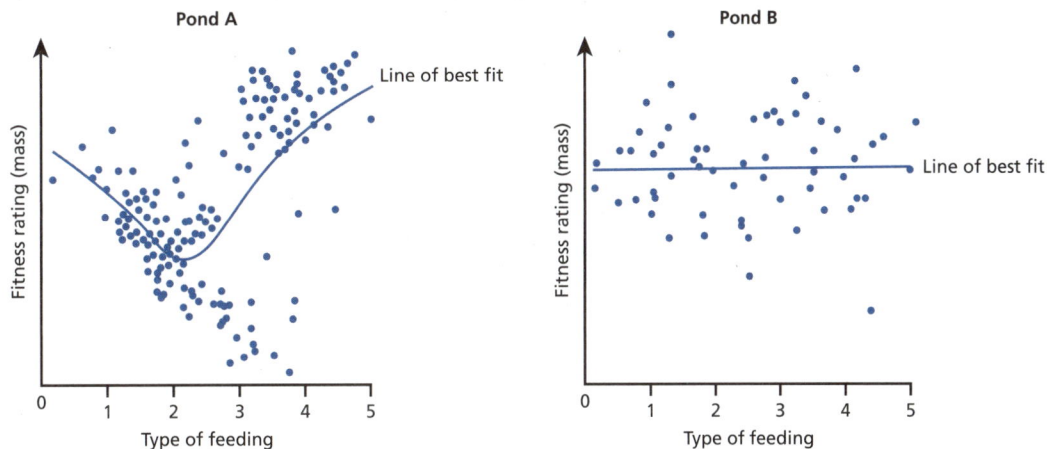

i) Identify if selection is occurring in each of the ponds. If it is occurring, state what type of selection it is and which tadpoles are being selected against. [3]

Pond A

Pond B

ii) The scientists noticed that in ponds with a large number of tadpoles, the results were similar to the results from pond A, whereas in ponds with a smaller population, the results were similar to those of pond B.

Suggest an explanation for this observation. [2]

iii) Describe what would need to happen for the toads living in a pond to develop into two different species. [3]

Types of speciation

Key words

Key questions

Why are islands often the site of allopatric speciation?

Why is genetic drift more likely to have an effect in small populations?

Why might differences in courtship displays lead to sympatric speciation?

Allopatric speciation

Isolating mechanisms are required to prevent two populations of a species from mating. This is needed to prevent gene flow and to allow speciation.

One type of isolating mechanism is seen in **allopatric speciation**. Allopatric speciation occurs when members of a species become separated into geographically isolated populations that can no longer interbreed. This can happen when a physical barrier develops, e.g. a new river or the movement of land masses by plate tectonics. This means that the pool of alleles in each population is isolated. The two different geographical areas may have different environmental conditions and so different selection pressures. Different mutations will be selected for in the two areas and so the gene pools will change. If, when reunited, the members of the populations can no longer breed to produce fertile offspring then they will have become new species.

The development of the 17 different species of finch in the Galápagos Islands (see page 118) is an example of allopatric isolation, as the separate ancestral populations were separated by the sea. **Genetic drift** may also contribute to allopatric speciation if the populations are small. This is a random process that means that some individuals may not pass on their alleles by chance and not due to selection. In small populations it may cause some alleles to completely disappear from the gene pool.

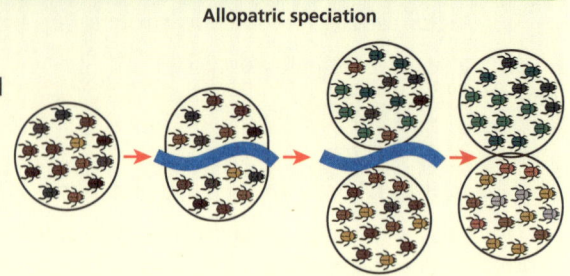

Allopatric speciation

Sympatric speciation

Sympatric speciation occurs when new species evolve from a single ancestral species in the same geographic region.

Although the organisms are all in the same area, some factor will prevent the two groups from mating. Examples of possible factors include differences in:
- courtship displays
- genitals in animals or in plant flower structure
- pheromones
- feeding preferences
- levels of activity throughout the day and night
- chromosome numbers.

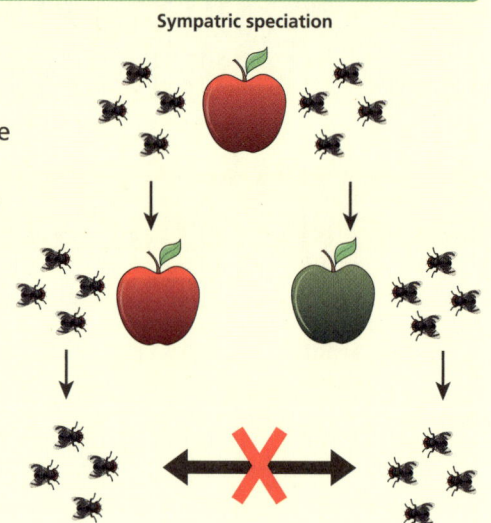

Sympatric speciation

An example shown in the diagram is the apple maggot fly in the USA. Apple maggot flies used to lay eggs on hawthorn, a tree native to America. When the domestic apple was introduced, some apple maggot fly females laid their eggs on the apples. Female apple maggot flies lay their eggs on the type of fruit they grew up in, so there were then separate populations growing up on the two different types of fruit. Over time, significant differences occurred, leading to the formation of new species.

Summary

Types of speciation

1. Lake Malawi is a large lake in Africa. It contains nearly 1000 different species of cichlid fish. The fish have all descended from a small number of fish that swam into the lake about 1 million years ago. Scientists have put forward two theories to explain how the large number of different species evolved.

Theory 1: As the level of water in the lake decreased, many different habitats were created and different species evolved in each habitat.

Theory 2: In each habitat in the lake, there was increased competition for food. Different jaw structures evolved within each habitat to allow feeding on different foods.

Which row of the table shows the type of speciation that is described by each theory? .... [1]

	Theory 1	Theory 2
A	sympatric	sympatric
B	allopatric	allopatric
C	sympatric	allopatric
D	allopatric	sympatric

2. Scientists have studied the mosquitoes that live in the tunnels where underground trains run in London. Populations of mosquitoes were trapped in the tunnels when they were built over 100 years ago.

The scientists compared the underground mosquitoes with populations that still live above ground. The table shows some of their observations.

Characteristic	Above-ground mosquitoes	Underground mosquitoes
Egg production	Only happens after they feed on blood from birds	Do not need a blood meal
Life cycle	Hibernate during winter	Do not hibernate

a) Explain how the underground mosquitoes have become adapted to their habitat. [3]

....

....

b) The scientists think that the underground mosquitoes may have formed a new species.

 i) Describe how they could test this theory. [2]

....

....

....

 ii) If the scientists are correct, what type of speciation has occurred? .... [1]

3. Anoles are lizards that live in tropical forests.

Scientists studied a species of anole that all live in the same forest. They found that some of the lizards spent more time in the trees and others spent more time on the ground. They found that the two groups of anoles have different mean leg lengths. The anoles that spent more time in the trees had significantly shorter legs.

a) Suggest a reason for the difference in mean leg length. [2]

....

....

b) Name the type of selection occurring in this population. .... [1]

c) If the two groups of anoles become different species, which type of speciation will have occurred? [1]

....

Populations in ecosystems

Key words

Key questions

What is a niche?

Why is intraspecific competition often greater than interspecific?

Why is the method of marking important in capture-recapture?

Components of an ecosystem

A **community** is all the living organisms of different populations in a habitat. The community and the **abiotic** (non-living) parts of the environment make up an **ecosystem**. Changes to the **biotic** (living) and abiotic components of an ecosystem mean that its makeup changes over time. The way an organism fits into an ecosystem is called its **niche**. The niche describes its interactions with the environment and with other organisms.

The size of a population depends upon limiting factors:
- In plants this could be herbivores, temperature, availability of light, water or minerals.
- For animals this could be food availability, predation or disease.

The **carrying capacity** is the maximum size that a population can reach and sustain indefinitely in a particular habitat. This will often be affected by competition for the resources.

Competition can be within a species (**intraspecific**) or between species (**interspecific**).

Gause's law of competitive exclusion states that two species that have the same niche cannot co-exist. Interspecific competition will lead to the eventual reduction in population size of one species.

This has been demonstrated in two species of *Paramecium* when they are grown separately or grown together.

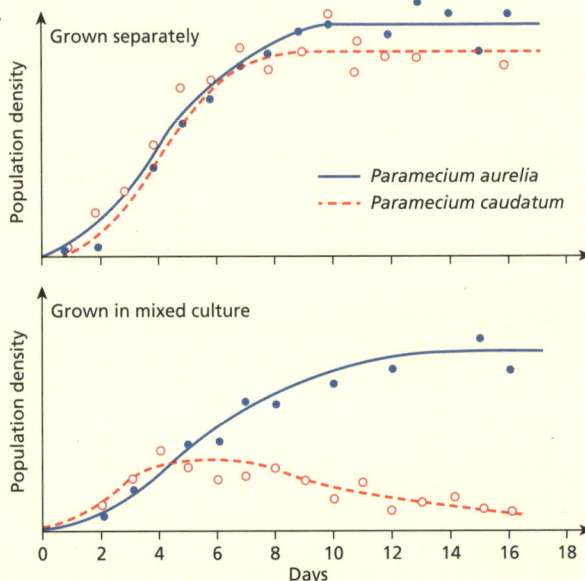

Grown separately

Population density

— *Paramecium aurelia*
--- *Paramecium caudatum*

Grown in mixed culture

Population density

Days

Investigating populations

The size of a population can be estimated:
- **Random quadrats** can be used to sample plants or slow moving animals (see page 78).
- **Belt transects** with quadrats can provide information on population size but can also be used to study how the distribution of organisms varies across an area.
- **Mark-release-recapture** (capture-recapture) can be used on motile animals.
  - organisms are captured, unharmed
  - they are counted, then marked in some way
  - they are released and, after some time, organisms are captured again
  - the number marked and unmarked are counted
  - this formula is then used to estimate the population:

Belt transect

$$\text{estimate of population size} = \frac{\text{number in first sample} \times \text{number in second sample}}{\text{number of marked individuals in second sample}}$$

The use of mark-release-capture assumes that there has been no migration in or out of the population, no births or deaths, the marking does not affect the organisms' survival, and all organisms have an equal chance of being captured.

Summary

Populations in ecosystems

1. Two species of mammal both spend time in a large field and in a woodland. In the summer and autumn, they both eat seeds from the grass in the field. In the winter and spring, one species eats acorns from the forest but the other eats roots in the field.

   a) Which statements are correct for these two species of mammal? Tick **one** box. [1]

   1. They are members of the same population.

   2. They occupy the same niche.

   3. They are members of the same trophic level.

   4. They are part of the same community.

   2, 3 and 4 only ☐   3 and 4 only ☐   2 and 3 only ☐   1 and 4 only ☐

   b) In the summer and autumn, the two species are competing with each other.

   Which row of the table summarises this competition? .... [1]

	Type of competition	Type of factor competing for
A	intraspecific	biotic
B	intraspecific	abiotic
C	interspecific	biotic
D	interspecific	abiotic

2. Scientists investigated the size of a population of deer in a very large habitat from 1962 to 1976. They also estimated the availability of grass in the same habitat. The results are shown in the graph.

   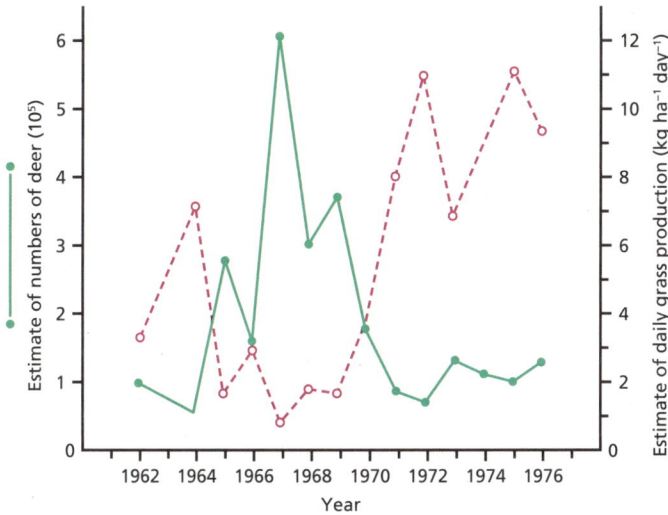

   a) Estimate the mean carrying capacity of the habitat for deer between 1972 and 1976. [1]

   ....

   b) Suggest an explanation for the variation in the deer numbers shown in the graph. [3]

   ....
   ....
   ....
   ....
   ....
   ....

   c) Explain how the graph would change if another herbivore was introduced into the habitat. [2]

   ....
   ....

3. Scientists use the capture-recapture technique to estimate the number of moths in a habitat.

   They capture 40 moths, mark them with a small dot of paint under their wings and then release them. The next day, they sample again and capture 30 moths. The scientists used these results to produce an estimate of 120 moths in the habitat.

   a) Calculate the number of moths in the second sample that were marked. [2]

   ....

   b) Explain what effect it could have on the population estimate if the scientists marked the moths with a larger dot on the top of their wings. [2]

   ....
   ....

Succession and conservation

Key words

Succession

Any area of land that is free from any organisms can become colonised. There will be a gradual change in the organisms living there over time. This change is called **succession**. If there is no soil initially present in the area, it is called a **primary succession**. If there is soil present, e.g. after a forest fire, then it is **secondary succession**.

The primary succession process goes through stages:

- The first organisms to colonise barren rock are called **pioneer species**, typically lichens and mosses that can grow without soil.
- Material from erosion and organic matter from the pioneer species forms soil, making the habitat more suitable for other species to live, which will then replace the pioneers.
- This process continues, with species changing the environment and making it more suitable for other species until a stable situation with a **climax community** is formed.

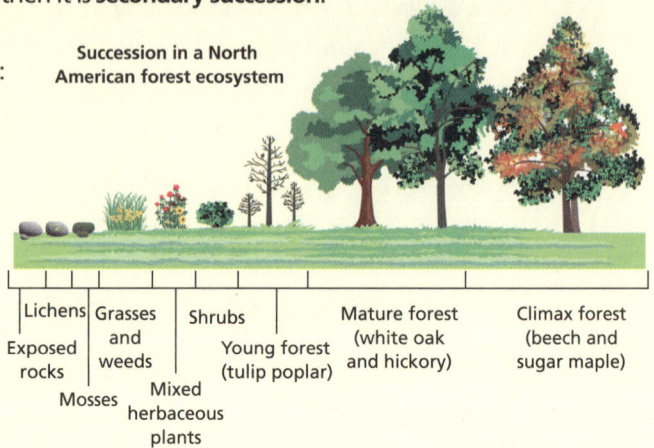

Each stage in the succession is called a **seral stage**. The climax community that is formed depends on a number of factors such as climate and soil condition. In many areas of Britain, this is deciduous woodland but further north it may be coniferous woodland.

Succession in a North American forest ecosystem

Exposed rocks | Lichens | Grasses and weeds | Mosses | Mixed herbaceous plants | Shrubs | Young forest (tulip poplar) | Mature forest (white oak and hickory) | Climax forest (beech and sugar maple)

Key questions

Why do seral stages contain different organisms?

What is the climax community in many tropical areas?

What was the original climax community in grassland areas of Britain?

Conservation

Conservation involves managing natural resources sustainably. Owing to human needs, resources often need to be removed from habitats, but steps should be taken to ensure they are replaced or replenished.

Conservation often involves manipulating succession. Examples include:

- **Management of forests** – a cycle of cutting down mature trees for their timber and planting new trees that will eventually be cut down decades later.
- **Maintenance of grasslands** by introducing animals, such as sheep and cows, that prevent shrub and tree growth. If the herbivores were removed, the succession would continue and climax forest would eventually grow.

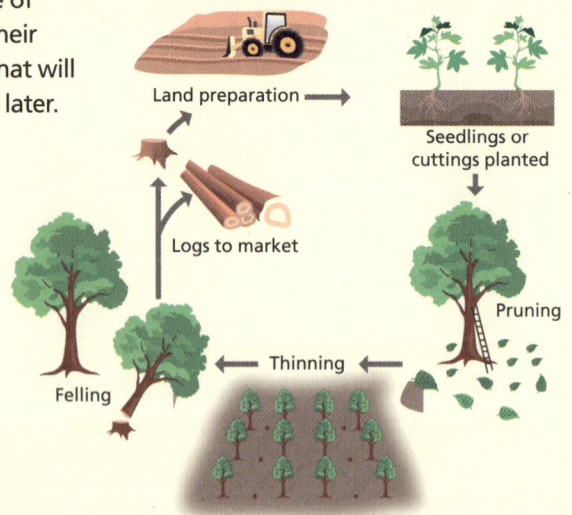

Many areas of Britain, such as grasslands, are thought of as being natural, but are actually managed. Many people think they should be conserved, as they have unique communities of organisms.

Human action to alter natural progression of succession results in a **deflected succession**.

Land preparation → Seedlings or cuttings planted → Pruning → Thinning → Felling → Logs to market →

Summary

Succession and conservation

1. Tick **one** box to show which statement is true for succession. [1]

   The climax community always has greatest biodiversity. ☐

   The environment of the ecosystem is constant over time. ☐

   The climax community that is formed can vary in different climates. ☐

   The pioneer species have the most complex soil requirements. ☐

2. Trees are often cut down for timber.

   The roots are dug up and new trees replanted. It can take over 50 years for the new trees to regrow. In some countries, mudslides have occurred during heavy rain because the soil is not held together by roots.

   Another technique used in woodlands is called coppicing. This harvests small areas of woodland at a time using the method shown in the diagram.

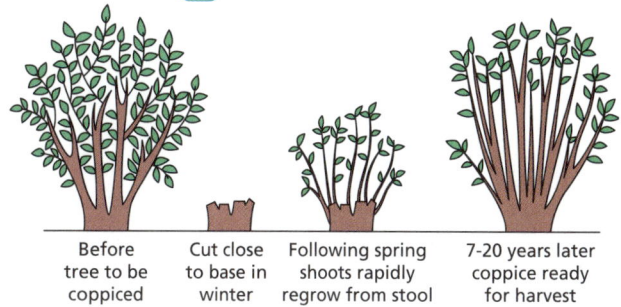

   | Before tree to be coppiced | Cut close to base in winter | Following spring shoots rapidly regrow from stool | 7-20 years later coppice ready for harvest |

   a) Suggest possible advantages and disadvantages of coppicing for harvesting timber. [4]

   ....

   ....

   ....

   b) Woodland that is used for recreation and not for timber is often coppiced. Explain how this increases the biodiversity of plants and animals on the woodland floor. [3]

   ....

   ....

3. Succession has occurred in a sand dune habitat. Graph 1 shows the change in the species living there over time. Graph 2 shows the change in the percentage of sunlight reaching ground level.

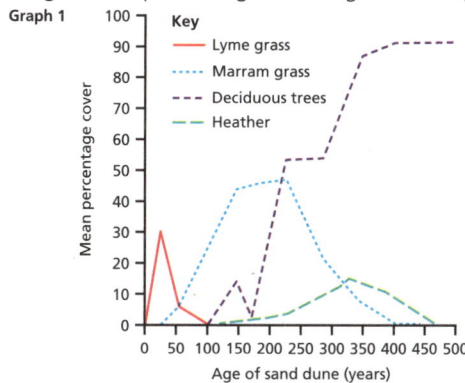

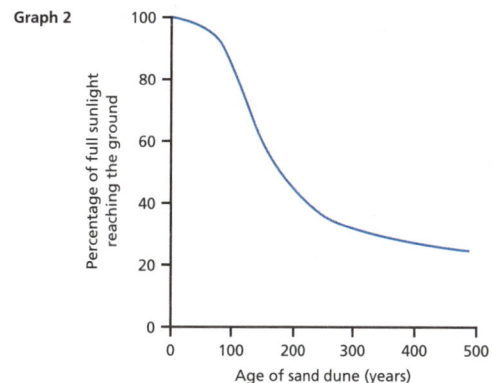

   Graph 1

   Key
   — Lyme grass
   ···· Marram grass
   -- Deciduous trees
   – – Heather

   Mean percentage cover / Age of sand dune (years)

   Graph 2

   Percentage of full sunlight reaching the ground / Age of sand dune (years)

   a) Name the pioneer species and the dominant species in the climax community. [2]

   Pioneer species = ...

   Dominant species in the climax community = ...

   b) Describe how the pioneer species changed the sandy habitat. [1]

   ....

   c) Use graph 1 to explain the pattern seen in graph 2. [2]

   ....

   d) Studies show that the highest net productivity of an ecosystem often occurs before the climax community is achieved. Suggest why this might be the case in this habitat. [2]

   ....

Mutations and protein structure

Key words

Key questions

At what stage of the cell cycle do mutations occur?

What does the term 'degenerate' mean?

Why is the introduction of a stop codon so harmful?

Types of mutation

Mutations occur during DNA replication (see page 68). This can be prior to meiosis or mitosis. The rate of mutation of DNA varies between species and even between genes in one species. An average mutation rate is between 10^{-7} and 10^{-8} per nucleotide, per cell division. This is an important figure: most mutations are harmful, so a higher rate would be a disadvantage to an individual, or even lethal; if the rate was too low, then new alleles would not be formed and the rate of evolution would slow.

The rate of mutation can be affected by the environment. Mutagenic agents (see page 68) can increase the mutation rate and can be physical, chemical or biological:

- Physical mutagens include high energy radiation (e.g. ultraviolet, gamma) which can form free radicals in cells, resulting in bonds forming between the wrong bases.
- Chemical mutagens include substances such as nitrosamines found in tobacco smoke, which can chemically react with bases, affecting pairing.
- Biological agents include certain viruses that can insert DNA into chromosomes.

Gene mutations causing changes to single genes include additions, substitutions and deletions (see page 68). There are other types of mutation that can be classed as gene mutations or chromosome mutations depending on the length of DNA involved:

- **Inversions** occur when a section of DNA is broken at two points, rotated by 180° and then rejoined.
- **Duplications** occur when a section of DNA is copied and repeated next to the original section.
- **Translocations** are when a piece of one chromosome breaks off and attaches to another chromosome.

Effects on protein structure

In a substitution mutation, one base is swapped with another base. In some cases, because of the degenerate nature of the genetic code, this will not cause a change in the amino acid coded for. However, an incorrect amino acid could be coded for and incorporated into the polypeptide chain. An example is seen in sickle-cell anaemia:

- Adenine is changed to thymine in the 6th codon of the gene coding for the beta chain of haemoglobin.
- This codes for the amino acid valine rather than the hydrophilic glutamic acid.

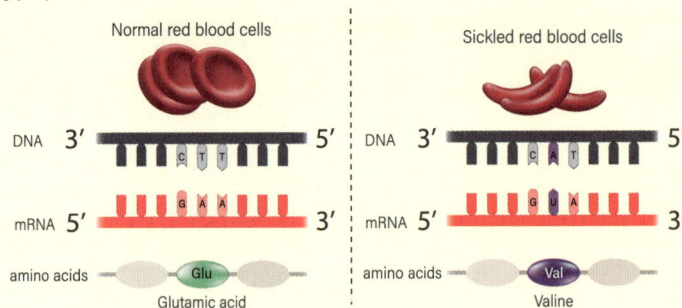

Normal red blood cells

DNA 3' ⋯ C T T ⋯ 5'
mRNA 5' ⋯ G A A ⋯ 3'
amino acids — Glu — Glutamic acid

Sickled red blood cells

DNA 3' ⋯ C A T ⋯ 5'
mRNA 5' ⋯ G U A ⋯ 3'
amino acids — Val — Valine

- As valine is hydrophobic, the tertiary shape of the molecule changes.
- This results in a change in the shape of red blood cells.

Mutations such as additions or deletions are more likely to have effects on the phenotype if the bases added or removed are not in multiples of three. This is because they can form frameshift mutations (see page 68).

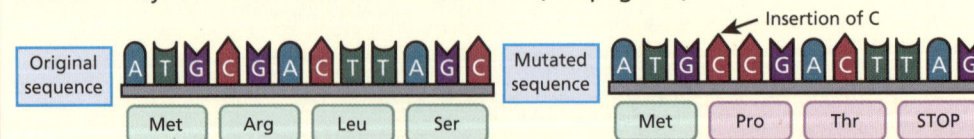

Insertion of C

| Original sequence | ATGCGACTTAGC |
| Met | Arg | Leu | Ser |

| Mutated sequence | ATGCCGACTTAG |
| Met | Pro | Thr | STOP |

Summary

Mutations and protein structure

① A gene codes for an enzyme that repairs faulty DNA.

The table shows the mRNA codons for seven different amino acids that are found in the enzyme.

a) One mutation in the gene changes the DNA triplet AAT to AAA.

   i) Explain what mutations could have caused this change.

....

....

....

Amino acid	mRNA codons
Asn	AAU, AAC
Cys	UGU, UGC
Gin	CAA, CAG
Leu	CUU, CUC, CUA, CUG, UUA, UUG
Lys	AAA, AAG
Met	AUG
Phe	UUU, UUC

[3]

   ii) Explain which amino acid is now coded for by this triplet. [2]

....

....

b) This is the DNA sequence of another section of this gene: AAGTTTGTT

   A different mutation causes a change to this DNA base sequence: AAGGTTGTT

   i) Explain what change this would make to the amino acid sequence. [1]

....

   ii) Explain how this could prevent this enzyme from working. [2]

....

....

   iii) Here is a third mutation to this gene: AAGGTGTT

   Explain why this third mutation is more likely to have a serious effect on the functioning of the enzyme. [3]

....

....

....

② The diagram shows the central region of an enzyme molecule with the amino acids numbered. Scientists wish to produce a mutated form of the enzyme, which is more stable at higher temperatures.

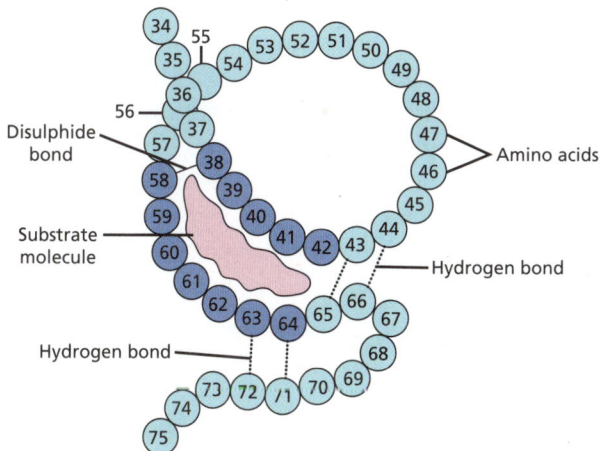

a) Amino acid 40 is a hydrophobic amino acid.

Explain why it would not be useful to produce a mutated enzyme with a hydrophilic amino acid at that position. [3]

....

....

....

b) Amino acids 38 and 58 are both cysteine and are coded for by the UGU codon.

Amino acids 43 and 65 are both serine amino acids and are coded for by UCU codons.

i) Explain what type of mutation would be needed to convert amino acids 43 and 65 to cysteine. [2]

....

....

ii) Explain why this mutation would make the enzyme more stable at high temperatures. [3]

....

....

Stem cells and specialisation

Key words

Key questions

What is the difference between totipotent and pluripotent stem cells?

What happens to the genes of a cell when the cell becomes specialised?

What is the function of transcription factors in producing iPS?

Types of stem cell

Stem cells are **undifferentiated** and can form specific, specialised cells. There are different types of stem cell:

Totipotent stem cells can become any type of cell and so could be used to grow a new human being. A zygote is the first stem cell of a developing body and is totipotent. The first few cells formed by the first two cell divisions are also totipotent. When a totipotent cell becomes specialised, only some of the genes are used and translated into proteins.

Pluripotent stem cells are also embryonic stem cells and can develop into any cell in the body but not cells in the placenta.

Multipotent stem cells can be found in adult tissues, can develop into more than one type of cell but are more limited than pluripotent stem cells. For example, haematopoietic stem cells in bone marrow can only form different types of blood cell – red, white and platelets.

Unipotent stem cells are more limited than multipotent cells as they can only produce one cell type. For example, cardiomyocytes (cardiac muscle cells) and skeletal muscle cells are produced by different unipotent stem cells.

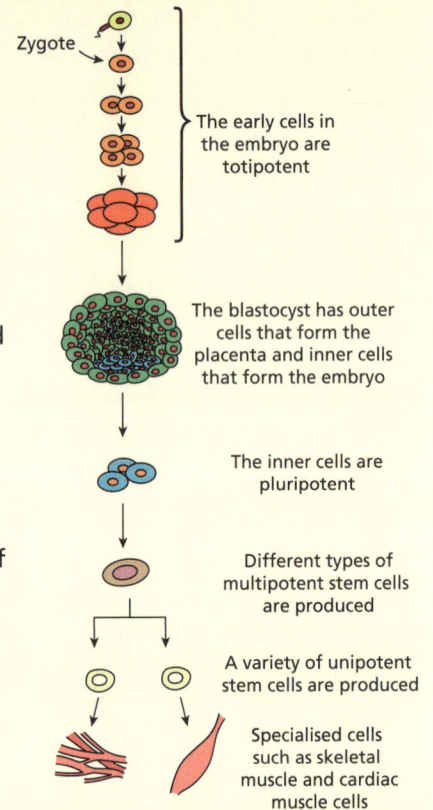

Zygote

The early cells in the embryo are totipotent

The blastocyst has outer cells that form the placenta and inner cells that form the embryo

The inner cells are pluripotent

Different types of multipotent stem cells are produced

A variety of unipotent stem cells are produced

Specialised cells such as skeletal muscle and cardiac muscle cells

Medical uses of stem cells

The ability of stem cells to differentiate means that they can be used for treating certain diseases and injuries. Pluripotent embryonic stem cells from early embryos have been used as they can become any cell in the body. These embryos have been produced by in vitro fertilisation (IVF) but have not been used for implantation. These stem cells have been used in a variety of ways:

- **Understanding why diseases develop** – stem cells can be grown in labs for scientists to better understand how human bodies grow and develop, and how diseases occur.
- **Testing** – researchers test drugs on stem cells to see if they are effective and safe.
- **Cell-based therapy** – involves treating damaged or diseased tissues or organs. Examples include treatments for Parkinson's disease, Alzheimer's disease, spinal injuries, heart disease, diabetes and skin burns. If the stem cell comes from an embryo cloned from the patient there will be no organ rejection issues. There are ethical issues with using embryonic stem cells from human embryos because the embryos are destroyed.

Scientists are now turning somatic (body) cells into pluripotent cells called **induced pluripotent stem cells** (iPS). They are made from somatic (body) cells, such as skin or blood cells, by reactivating genes that have been turned off when the cells became specialised. This is done using protein transcription factors.

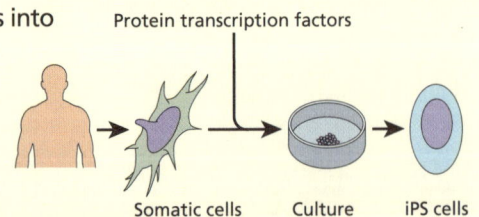

Protein transcription factors

Somatic cells   Culture   iPS cells

Summary

Stem cells and specialisation

1 Which of these is correct for induced pluripotent stem cells? Tick **one** box. [1]

They are made from embryos left over from IVF. ☐   They can form the placenta. ☐

They are produced from body cells. ☐   They can form only one type of tissue. ☐

2 Complete this table by adding ticks (✔) or crosses (✘) to show the features of different types of stem cell. [4]

Feature	Type of stem cell		
	Totipotent	Pluripotent	Multipotent
Can be isolated from adult tissues			
Can form placental cells			
Have some of their genes switched off			
Can form all the cells in the human body			

3 Mesenchymal stem cells are found in the tissues of the skeleton, such as bone, muscle and cartilage. They can differentiate into these three types of cell.

- **Graph 1** shows how the percentage of bone marrow cells that are mesenchymal stem cells changes with age.

- **Graph 2** shows the percentage of bone fractures that healed after one year in different aged patients.

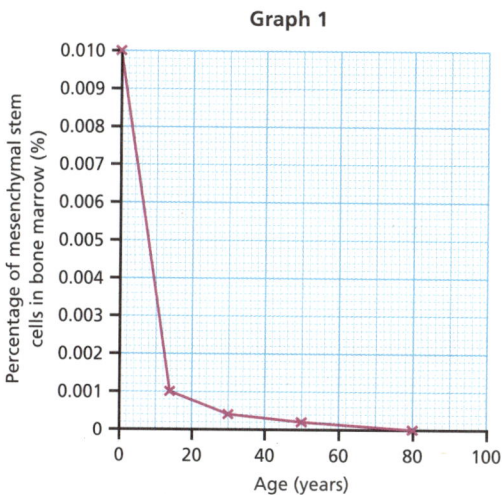

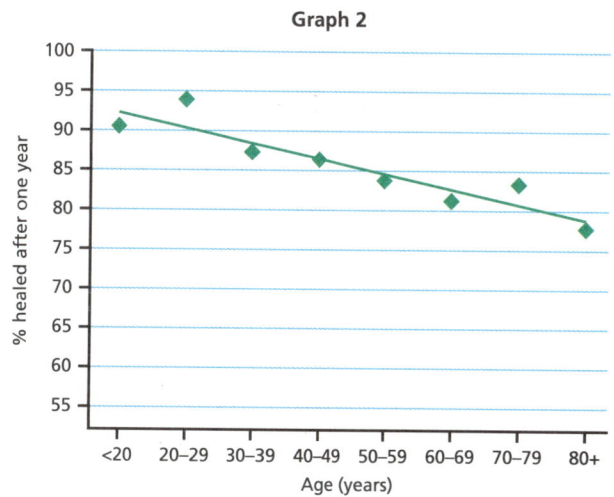

Graph 1

Graph 2

a) State the type of stem cell that includes mesenchymal stem cells. .... [1]

b) Use **graph 1** to calculate the mean rate of change in the percentage of mesenchymal stem cells between the ages of 10 and 40.

.... [3]

c) Use the data in **graph 1** to explain the pattern of results shown in **graph 2**. [3]

d) Mesenchymal stem cells can be easily isolated from adult bone marrow and are the most commonly used stem cells in treating patients.

Discuss the advantages and disadvantages of using mesenchymal stem cells rather than embryonic stem cells for treating patients. [4]

Control of gene expression

Key words

Key questions

Are the responses caused by oestrogen immediate or longer term?

If histones prevent DNA unwinding, what effect will this have?

How does siRNA bind with single-stranded mRNA?

Stimulating transcription

All body cells in eukaryotes contain the same genes, but when cells become specialised only some of these genes are translated to form proteins. One method of controlling gene expression is through **transcription factors**. They:

- are proteins that bind to specific DNA sequences to control the rate of transcription from DNA to mRNA
- can work on their own or as a complex with other proteins to promote the attaching or the blocking of RNA polymerase.

Steroid hormones, such as oestrogen, can switch on a gene and start transcription:

- Oestrogen can pass through the cell membrane and enter cells.
- The oestrogen then binds to a receptor combined with a transcription factor.
- The transcription factor changes shape and can then bind to a promoter region of the DNA, starting transcription.

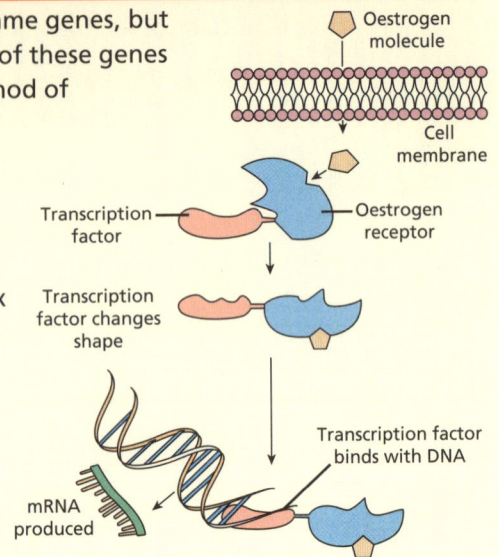

Epigenetics

Epigenetics is the study of the changes in organisms caused by modifying gene expression rather than changing the DNA base sequence. These modifications can be triggered by changes in the environment and can either enable transcription or turn off a gene. It is thought that some of these changes can be inherited.

There are two main ways that epigenetics can occur:

- **DNA methylation** – methyl groups are added to DNA. The only bases that can be methylated are cytosine and adenine. Adenine methylation only occurs in prokaryotes. Methylation alters DNA function, typically repressing transcription. DNA methylation occurs naturally during development, including in embryo development and aging.

- **Histone acetylation** – histones are eukaryotic proteins that package DNA into nucleosomes. When a histone is acetylated by the addition of acetyl CoA, the gene is transcribed. When the histones are deacetylated, transcription is stopped.

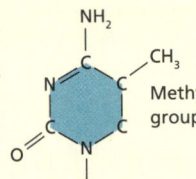

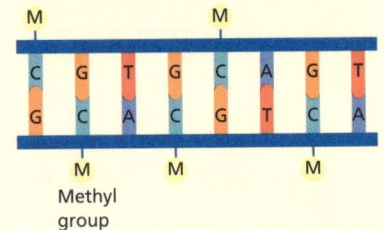

RNA interference

In eukaryotes and some prokaryotes, **RNA interference** (RNAi) can occur. This is a way of controlling which genes are translated after transcription has occurred. Short, double-stranded lengths of RNA called siRNA can split to become single-stranded. They can then bind with mRNA, making part of it double-stranded. This prevents the mRNA attaching to ribosomes and being translated. The double-stranded mRNA is destroyed by an enzyme complex called RISC. This process is called **gene silencing.** Cells can use it to provide protection from viruses that may have injected their genetic material. Scientists can use it as a method of blocking gene expression for genes that may cause disorders.

Summary

Control of gene expression

1) These are steps in the initiation of transcription by oestrogen. They are not in the correct order.

    A. The transcription factor changes shape.

    B. mRNA is formed.

    C. RNA polymerase can attach to DNA.

    D. The transcription factor binds to the promoter site on the DNA.

    E. Oestrogen binds with a receptor site.

    F. Oestrogen diffuses through the cell membrane.

a) Give the correct order for these steps by writing the letters in the boxes. [5]

☐ ☐ ☐ ☐ ☐ ☐

b) Explain why oestrogen is able to pass through the cell membrane, but hormones such as insulin cannot. [2]

c) Hormones such as insulin, glucagon and adrenaline work by activating existing enzymes.

    i) Explain how the action of oestrogen differs to the action of these hormones. [1]

    ii) Explain why the mechanism used by oestrogen would not be suitable for producing the type of response triggered by adrenaline. [3]

2) Complete the table with ticks (✔) or crosses (✗) to show the features of processes that can alter gene expression. [4]

Feature	Process		
	Use of transcription factors	Epigenetics	RNA interference
Destroys RNA polymerase			
Involves enzymes			
mRNA is destroyed			
Involves methylation			

3) The diagram shows the process of gene silencing by siRNA.

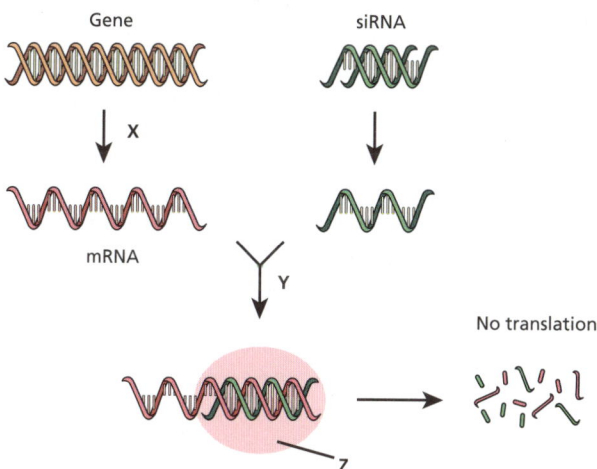

Gene

siRNA

X

mRNA

Y

No translation

Z

a) Name the process occurring at X. [1]

b) Explain what is occurring at Y. [2]

c) Name molecule Z. [1]

d) Explain why this process is called gene silencing. [3]

e) Explain how this process could be used by doctors to prevent viruses damaging cells. [3]

Gene expression and cancer

Key words

Key questions

Why might some benign brain tumours be difficult to remove?

What might turn on proto-oncogenes at the wrong time?

Which type of genes are likely to be turned on in the development of breast cancer?

What is cancer?

Cancer is a condition where cells in a specific part of the body grow and reproduce uncontrollably. When cancer cells produce a lump or a growth, it is called a **tumour**. There are two main types of tumours:

Benign tumour   Malignant tumour

- **Benign tumours**: These are often called non-cancerous tumours. This is because the cells do not spread to other parts of the body. Although they multiply faster than normal cells and do not die, the tumour grows fairly slowly. This can cause problems because it may press on surrounding tissues.
- **Malignant tumours:** These tumours usually grow faster than benign tumours. They can spread cancer cells to other parts of the body via the bloodstream or lymphatic system. This spread is called **metastasis** and secondary tumours can form. These can cause damage or possibly death.

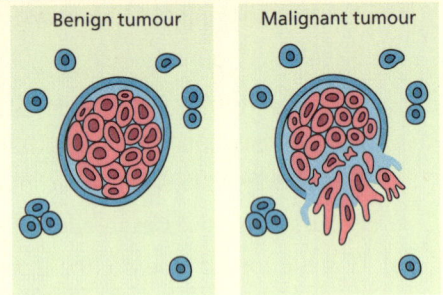

Two types of genes play a role in the development of cancer:

- **Proto-oncogenes** are genes that normally stimulate cells to grow and divide to make new cells.
- **Tumour suppressor genes** are genes that slow down cell division or trigger cells to die at the correct time **(apoptosis)**.

Mutations in either of these two types of gene can cause cancer. When a proto-oncogene mutates, it can become turned on (activated) when it is not supposed to be. It is now called an **oncogene**. This can cause the cell to grow and divide out of control, which might lead to cancer.

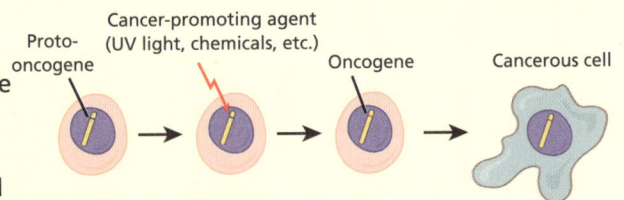

Proto-oncogene | Cancer-promoting agent (UV light, chemicals, etc.) | Oncogene | Cancerous cell

Mutations to tumour suppressor genes can remove any control of the cell cycle and cells may not die. This can also lead to cancer.

Abnormal methylation of oncogenes and tumour suppressor genes can be the cause of cancer.

Methylation of tumour suppressor genes may prevent them being transcribed and lack of methylation can turn on oncogenes.

Some types of breast cancer can be triggered by high levels of oestrogen. Oestrogen binds to receptors and the transcription factors initiate transcription. In some cases this may lead to cancer, although the exact mechanism is not fully understood.

Cancer treatment

By understanding the roles of oncogenes, tumour suppressor genes and the nature of epigenetics, scientists have developed new ways to treat and possibly cure cancer.

Previous treatments involved chemotherapy. This targeted rapidly dividing cells but also killed some healthy cells as well as the cancer cells, therefore causing side effects.

It is now possible to affect the epigenetic process by altering methylation. The activity of oncogenes and tumour suppressor genes can therefore be modified.

✔ Summary

Gene expression and cancer

1. Tick **one** box to show which statements are correct about tumours. [1]

1. Benign and malignant tumours both contain cells that live longer than normal cells.

2. Benign tumours do not form metastases.

3. Malignant tumours usually grow faster than benign tumours.

1, 2, and 3 ☐   2 and 3 only ☐   1 and 3 only ☐   1 and 2 only ☐

2. The diagram shows one mechanism for the development of cancer. A gene called DNMT1 codes for the production of an enzyme called DNA methyltransferase. Overactivity of this gene can result in some types of cancer.

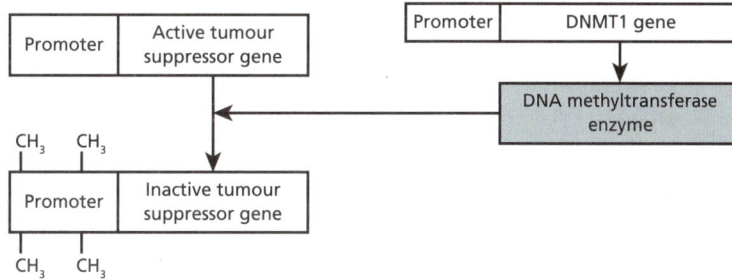

a) Name the type of gene that DNMT1 is acting as in the development of cancer. .... [1]

b) Use the diagram to explain how DNMT1 can cause cancer. [3]

....

....

....

....

c) A different gene called HDAC1 can also cause cancer. It codes for a different enzyme that does not affect DNA but removes acetyl groups from histones.

Suggest how this could cause cancer. [2]

....

....

3. The diagram shows how oestrogen is produced from androgens and how it affects cells in the breast.

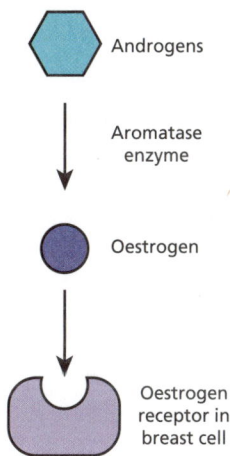

a) Aromatase inhibitors can be used to treat breast cancer in some women.
Explain why they work. [3]

....

....

....

....

b) Suggest a possible disadvantage of this type of treatment. [2]

....

....

c) Another drug called tamoxifen can be used as a treatment.

This diagram represents the shape of the tamoxifen molecule.

Suggest how tamoxifen treatment works. [3]

....

....

....

Amplifying DNA

Key words

Key questions

Why are sticky ends given that name?

How can scientists select where DNA is cut?

What bonds are produced by DNA polymerase?

Producing DNA fragments

Many types of gene technology involve the production of DNA fragments from longer sections of DNA or by using mRNA. These fragments can be produced in several ways:

- **Restriction enzymes (restriction endonucleases)** are used to cut a section of DNA or a gene from a donor organism.
- **Reverse transcriptase,** an enzyme originally discovered in retroviruses, can use mRNA to produce complementary DNA (cDNA).
- **Gene machines,** introduced in 2015, can automatically piece together DNA nucleotides in the desired order.

Different restriction endonucleases target a different base sequence and cut the DNA at that point, e.g. the enzyme EcoRI cuts the sequence GAATTC. However, the cuts are staggered, producing fragments with 'sticky' ends (ends made of single-stranded DNA). This is important as it provides a way of joining together fragments by complementary base pairing if required.

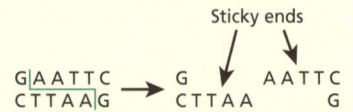

Amplifying DNA

Whichever method has been used to create DNA fragments, the DNA can then be amplified. This means producing many copies of the DNA. This is often achieved through a process called the **polymerase chain reaction (PCR)**. This is an *in vitro* method.

1. Denaturing – DNA is first heated to separate it into two strands.
2. Annealing – short lengths of DNA called primers are then added during the annealing phase.
3. Extension – a type of DNA polymerase that is not denatured by heat (called Taq polymerase) extends the new DNA chains through the addition of nucleotides.
4. New DNA molecules are produced and the process is repeated until there are millions of copies of the DNA.

The different stages of the PCR method are now fully automated in machines and can occur rapidly.

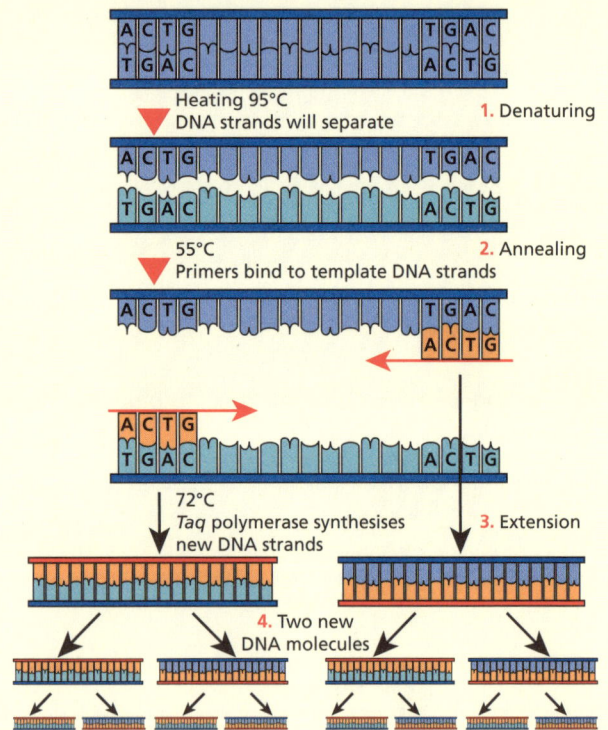

Summary

Amplifying DNA

(1) Tick **one** box to show which molecules are needed for the polymerase chain reaction. [1]

1. DNA polymerase
2. RNA polymerase
3. DNA primers
4. Amino acids

1, 2, 3 and 4 ☐   1 and 3 only ☐   2 and 3 only ☐   1 and 4 only ☐

(2) Why is Taq polymerase used in the polymerase chain reaction? Tick **one** box. [1]

It reduces the time needed to cool down the DNA from 95°C. ☐

It will join DNA nucleotides in both directions, from 5' to 3' and from 3' to 5'. ☐

It works at a range of pH values. ☐

It can join DNA nucleotides and join RNA nucleotides. ☐

(3) The diagram shows one cycle of the polymerase chain reaction (PCR).

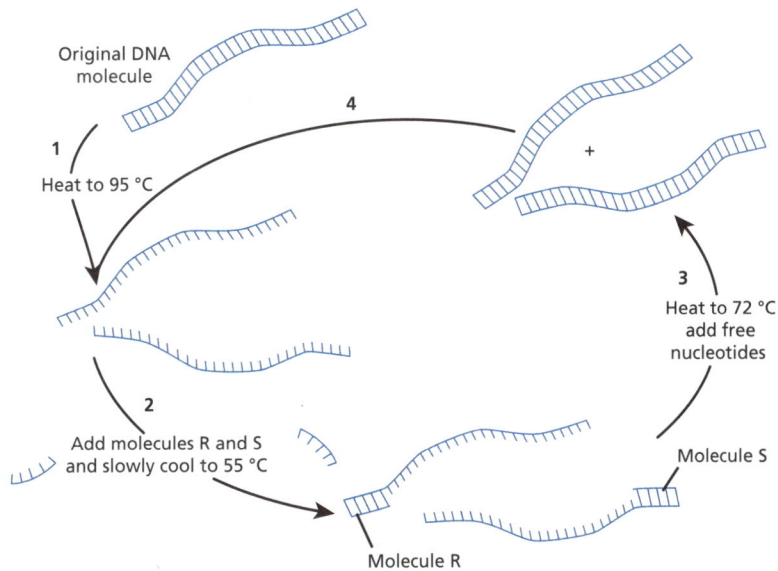

a) Name the short lengths of DNA molecules labelled R and S. .... [1]

b) Explain why two different molecules, R and S, are needed. [2]

....

....

....

c) Complete this table to show the reason why three different temperatures are used throughout the reaction. [4]

Temperature (°C)	Time at this temperature (s)	Reason for this temperature
95	30	
55	45	
72	60	

d) One DNA molecule is placed into the PCR machine.

Calculate how many molecules of DNA would be made in 18 minutes. [3]

Reading DNA

Key words

Key questions

When sequencing using a gel, which fragments travel further?

What bonds hold the probe to the strand being tested?

What causes the DNA fragments to be different lengths?

DNA sequencing

It is now possible to map the order of nucleotides in DNA molecules by **DNA sequencing**. This process was originally carried out using this method:

- A single-stranded length of DNA is cut into many fragments.
- A DNA primer and DNA polymerase produce a complementary strand to the sample.
- However, small quantities of four special nucleotides are used, called dideoxynucleoside triphosphates (ddNTPs).
- The ddNTPs are labelled in some way and are unable to form chemical links to any other nucleotides.
- Fragments of different sizes are therefore formed because the elongation is stopped when a ddNTP is incorporated.
- The different sized fragments are removed and separated by size.

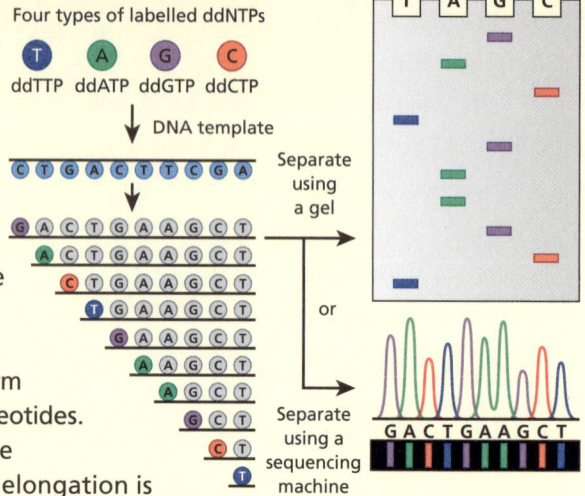

The detection of the different fragments was first achieved using radioactive markers. This allowed it to become fully automated, enabling the entire human genome to be sequenced by 2003. Now, new generation methods are used that are even faster, as many sequencing reactions happen at the same time.

DNA probes are fragments of DNA, which are radioactively or fluorescently labelled. They are used to detect specific nucleotide sequences in DNA that are complementary to the sequence in the probe. The single-stranded probe will join with the complementary base sequence on the DNA that is tested. This is called **hybridisation**. The specific DNA sequence can then be detected.

DNA profiles

It is now possible to produce a **genetic profile** to identify a specific person. It works because an organism's genome contains many **variable number tandem repeats (VNTRs)**. VNTRs are found in non-coding DNA and are short sequences of DNA that are repeated many times. The pattern and length of VNTRs vary in different individuals. The probability of two individuals having the same VNTRs is very low, so the genetic profile is unique.

1. DNA is extracted from the individual.
2. Different restriction enzymes cut the DNA into different lengths.
3. DNA fragments are separated according to length by **gel electrophoresis**.
4. The DNA is transferred to a membrane by blotting.
5. The membrane is then incubated with radioactive probes that will bind to any complementary DNA.
6. The DNA can be visualised by laying the membrane on a photographic plate.

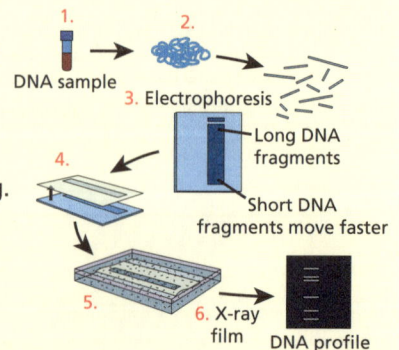

Summary

Reading DNA

1 Tick **one** box to show which of these molecules are used to sequence a length of DNA. [1]

   1. RNA polymerase

   2. ddNTPs

   3. RNA nucleotides

   4. DNA polymerase

   1, 2, and 4 only ☐   2, 3 and 4 only ☐   2 and 4 only ☐   1 and 4 only ☐

2 These are steps in the production of a DNA profile. They are **not** in the correct order.

   A. Separate fragments by gel electrophoresis.

   B. Cut the DNA into different lengths.

   C. Incubate with radioactive probes.

   D. Lay the membrane on a photographic plate.

   E. Transfer DNA to a membrane by blotting.

   F. Extract DNA from an individual.

   a) Give the correct order for these steps by writing the letters in the boxes. [5]

      ☐ ☐ ☐ ☐ ☐ ☐

   b) Explain the use of gel electrophoresis in the production of a profile. [2]

   ....

   ....

   c) Figure 1 shows a DNA profile and Figure 2 shows how the distance moved by DNA fragments in gel electrophoresis varies with their size.

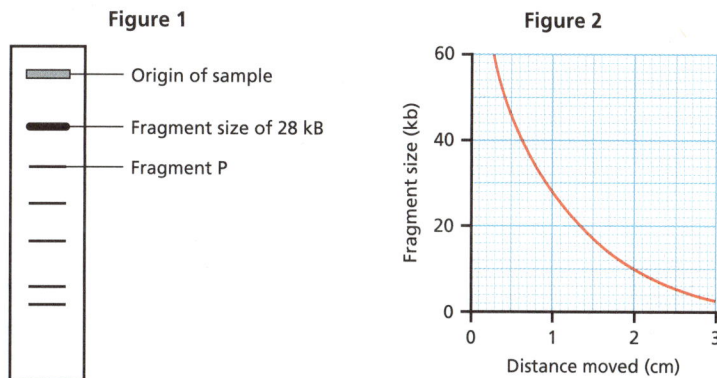

Figure 1

- Origin of sample
- Fragment size of 28 kB
- Fragment P

Figure 2

Fragment size (kb) vs Distance moved (cm)

   Use **figure 1** and **figure 2** to estimate the size of fragment P. .... [1]

   d) The speed at which DNA can be sequenced has been increasing rapidly since its introduction. When first developed, the speed at which DNA could be sequenced was approximately 500 base pairs per hour. By 2016, that speed had increased to approximately 50 million base pairs per hour.

      i) Calculate how many times faster the speed of DNA sequencing was in 2016 compared with when it was first developed. [2]

      ....

      ii) Give **one** way in which this increase in speed has been achieved. [1]

      ....

      ....

Recombinant DNA technology

Key words

Key questions

What bonds are formed by DNA ligase?

Why can some viruses be used as vectors in human cells?

Why are the bacteria grown on agar containing antibiotic?

Genetic modification

Genetic modification involves the transfer of fragments of DNA from one organism to another. Since the genetic code is universal, as are transcription and translation mechanisms, the transferred DNA can be translated within cells of the recipient organism. This organism is termed **transgenic** if it has received DNA from another species. The DNA that now contains material from another individual is called **recombinant DNA**.

The process of genetic modification in bacteria is shown in the diagram.

1. A restriction endonuclease cuts donor DNA and produces 'sticky ends'. Promoter and terminator regions must be included.
2. A plasmid is taken from a bacterium and cut with the same restriction endonuclease.
3. The gene is inserted into the plasmid by another enzyme known as DNA ligase.
4. The plasmid replicates inside the bacterium.
5. The bacteria reproduce and express the gene.

The bacteria make many copies of the gene and so this is an *in vivo* method of amplifying DNA.

The plasmid is described as a **vector** as it carries the recombinant DNA into the bacterium. Gentle heat treatment makes the bacteria take up the plasmid. This is called **transformation**.

A bacterium is often used to insert the gene into plant cells. Vectors for human cells can be viruses. If genes from eukaryotic cells are inserted into prokaryotic cells, the introns need to be removed first. This is done using the enzyme reverse transcriptase (see page 134). mRNA that has had the introns removed is converted to cDNA.

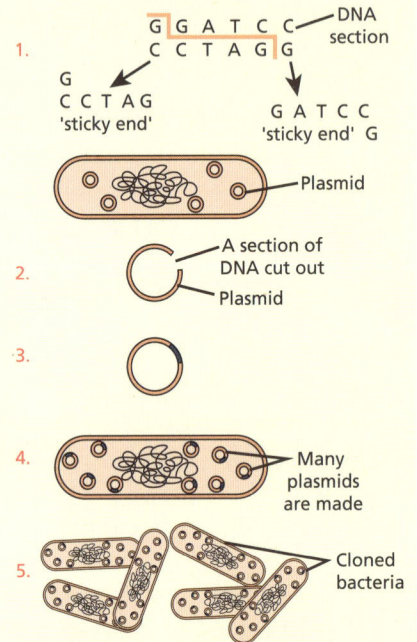

Use of marker genes

It is important to be able to tell which cells have taken up the recombinant DNA. This is achieved by adding an additional marker gene. There are two types of marker gene that are used:

- A **selectable marker** gives the host cells that possess the gene protection from a treatment that would normally kill it, e.g. antibiotic resistance.
- A **screenable marker** makes the host cells that have the gene appear different, e.g. fluorescing under UV light.

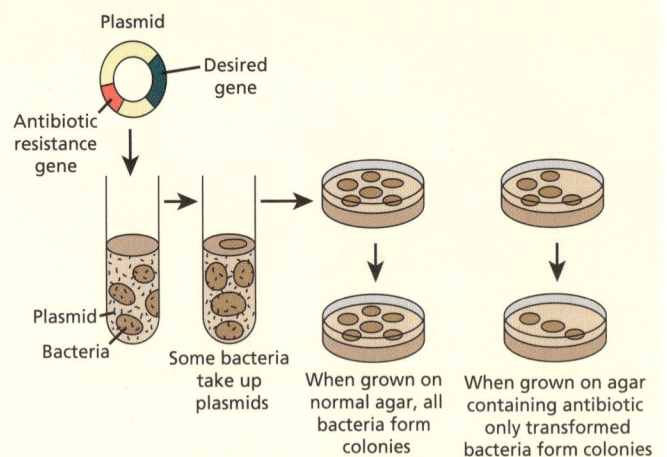

Summary

Recombinant DNA technology

1. These are steps in genetically modifying bacteria so that they express a human gene. They are **not** in the correct order.

   A. Identify the transformed bacteria.

   B. Use reverse transcriptase to produce cDNA.

   C. Use DNA ligase to join the DNA.

   D. Treat a plasmid with endonuclease.

   E. Obtain mRNA made from the human gene.

   F. Heat-treat a bacterial sample.

   a) Give the correct order for these steps by writing the letters in the boxes. [5]

      ☐ ☐ ☐ ☐ ☐ ☐

   b) Explain the importance of using reverse transcriptase in this process. [3]

2. Scientists want to produce a plant that is resistant to the weedkiller glyphosate. They have found a gene that provides resistance (gly^r).

   The first stage in producing a resistant plant is to insert this gene into a type of bacterium that can infect the plants and transfer the gene. The process for inserting this gene into the bacteria is shown in the diagram.

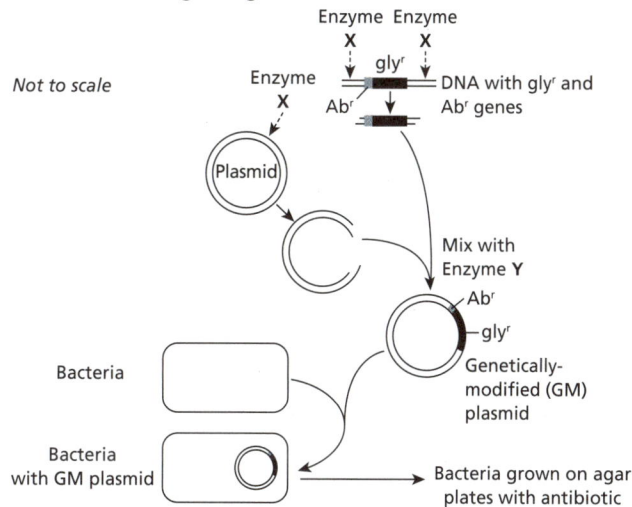

   a) Name enzyme **X** and describe its function in this process. [3]

   b) Explain why it is important to use the same variety of enzyme **X** on the donor DNA and on the plasmid. [2]

   c) Name enzyme **Y** and give its function in this process. [2]

   d) State how bacteria are treated so that they take up the GM plasmid. [1]

   e) State why the bacteria that have taken up the plasmid are described as transgenic. [1]

   f) Explain why the bacteria produced are grown on agar plates that contain antibiotic. [3]

Uses of DNA technology

🔓 Key words

❓ Key questions

What is a phylogenetic tree?

What is done to a very small sample of DNA to enable it to be fingerprinted?

How could the working gene be inserted into a person's cells?

Gene sequencing and DNA probes

Producing gene sequences and DNA profiles (see page 136) have many important uses.

Evolutionary biology
To map changes in organisms and to create phylogenetic trees. This allows the relationship of one species to another to be mapped (see page 74). If extinct organisms have DNA that can be analysed, the evolutionary relationships of extinct to living species can be investigated.

Genetics and immunology
It is possible to examine genotype–phenotype relationships, seeing how the genotype is expressed as the phenotype in living organisms. In simpler organisms, scientists can work out the amino acid sequences of all the proteins (the proteome) that the organism can produce. This can be important in pathogens for identifying antigens that might be useful for producing vaccines. In more complex organisms, the presence of non-coding DNA and regulatory genes means that the genome sequence cannot be so easily translated into the proteome.

Food production
Sequencing has allowed plant breeders to identify key genes and alleles that are important in features such as disease resistance, yield and drought tolerance. It can also protect the consumer, by testing products to see which organism a product actually comes from.

Medicine
The results can be used to identify causes of diseases and so develop ways for them to be treated. DNA probes can screen individuals for inherited conditions. This can provide information for genetic counselling. Some conditions are not predetermined by genes but the presence of certain alleles can give individuals an increased risk of developing a condition. Identifying these alleles will allow the individual to take preventative measures. There are also many variants of alleles that influence the effectiveness of drugs. Therefore, sequencing a person's DNA can allow personalised medicines to be developed. This is called **genomic medicine**.

Forensic science
Utilises DNA sequences to provide evidence for guilt or innocence. This is used in the production of DNA profiles or genetic fingerprints.

The VNTRs are identified from DNA samples using the technique described on page 136.

In the figure, DNA from a crime scene is compared with that of three suspects. The `fingerprint' that matches the one at the crime scene belongs to the culprit (Suspect 2).

There are a number of high-profile cases where DNA analysis of the historical evidence from the crime scene has proved that the convicted person could not have committed the crime.

DNA profiling is also used to help establish parentage by comparing the DNA sequence of a child with that of an adult to determine if they really are the parent.

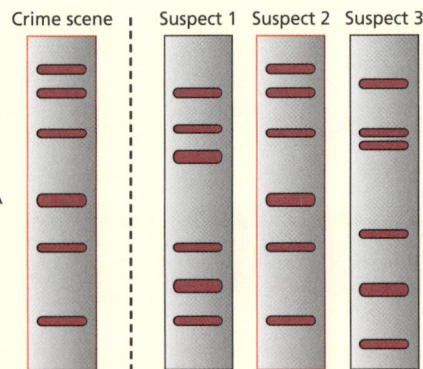

Crime scene   Suspect 1   Suspect 2   Suspect 3

Gene therapy

There is the potential to use recombinant DNA technology (see page 138) to help cure genetic disorders.

If a person has a defective gene that causes a genetic condition, it is possible to introduce a working copy of the gene into the person's cells. The working gene then produces the correct or missing polypeptide. This is called **gene therapy**. This type of gene therapy, changing the genes of body cells, is currently legal in the UK.

Changing the genes in the gametes is called **germ cell gene therapy**. This gene would appear in all the cells produced by the gamete and would also be passed on to future generations. This is not currently legal in the UK. Gene therapy, as with many of the applications of DNA technology, raises ethical, financial and social issues that should be debated before any application is used.

✓ Summary

Uses of DNA technology

1 Read the following passage about a possible use of genetically modified algae.

> Polyethylene terephthalate (PET) is a synthetic plastic. It has been widely used in agriculture and industry. It was considered to be non-biodegradable. Currently, it is dumped in landfill sites or burned. In 2016, a bacterium was found that produced PETase enzyme that hydrolysed PET.
>
> Scientists are now hoping to transfer the gene for PETase into photosynthetic algae.
>
> The advantages of using algae to produce PETase rather than bacteria are:
> * the low cost of growing the algae, as they do not need expensive food sources
> * antibiotic-resistant bacteria do not need to be produced to select the genetically modified organisms.

a) Suggest why scientists were keen to find a way of hydrolysing PET. [2]

b) Algae have cell walls like plant cells.

Suggest a vector that could be used to insert the PETase gene into algae. [1]

c) Give the reason why algae do not need expensive food sources to grow. [1]

d) Explain why scientists would prefer to avoid the production of antibiotic-resistant bacteria that are often made in some types of genetic modification. [2]

2 The diagram shows DNA profiles of six people that can be used for genetic fingerprinting.

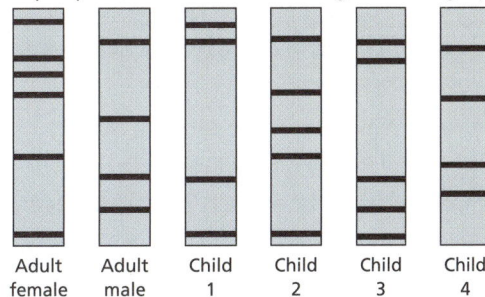

Adult female, Adult male, Child 1, Child 2, Child 3, Child 4

a) Genetic fingerprinting can be used to establish parentage.

Use the diagram to discuss what can be concluded about the parentage of the four children. [2]

b) Genetic fingerprinting can also be used to identify individuals from very small samples of DNA left at crime scenes.

i) Give the reason why PCR technology makes it possible to produce genetic fingerprints from very small samples. [1]

ii) Give the reason why genetic fingerprints are unique for individuals. [1]

3 Cystic fibrosis is a genetic disorder caused by a mutation in the CTFR gene. If both alleles for this gene are faulty, a person cannot produce a functional protein. Scientists have altered viruses so that they contain a working copy of the faulty gene. People with the condition can then inhale the viruses.

a) Genetic conditions that are caused by lack of a functional protein, rather than by the production of a harmful protein, are more likely to be successfully treated by gene therapy. Explain why. [3]

b) Scientists have found that people with a specific type of CTFR mutation, known as G551D, can be very effectively treated with a new drug called ivacaftor.

Give the name for this type of treatment. [1]

Required practical 1: Rate of enzyme-controlled reactions

Aim

To investigate the effect of a named variable on the rate of an enzyme-controlled reaction

Introduction

The effect of temperature on the rate of breakdown of hydrogen peroxide by the enzyme catalase can be investigated using potato tissue as a source of catalase. Catalase breaks down hydrogen peroxide to release oxygen gas. The rate of oxygen release is a measure of the rate of reaction.

Example method

1. Set the water bath to a temperature of 25°C.
2. Put 25 cm^3 of hydrogen peroxide solution into a boiling tube and leave the boiling tube in the water bath for 5 minutes.
3. Cut two cylinders of potato tissue with a cork borer.
4. Drop the potato cylinders into the boiling tube and connect the delivery tube.
5. Collect the oxygen for 2 minutes and measure its volume using the measuring cylinder.
6. Repeat the experiment at a range of different temperatures.

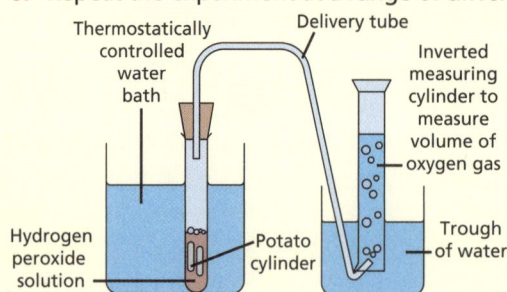

Independent variable	Temperature
Dependent variable	Volume of oxygen produced
Control variables	pH
	Concentration of hydrogen peroxide
	Size and number of potato cylinders

Analysis

Plot a graph of the volume of oxygen collected in 2 minutes against temperature.

The optimum temperature for catalase activity coincides with the peak of the curve.

The temperature coefficient (Q 10) is the factor by which the rate of the reaction increases for every 10-degree rise in the temperature. In this example, an increase from 20°C to 30°C doubles the rate from 3.0 to 6.0

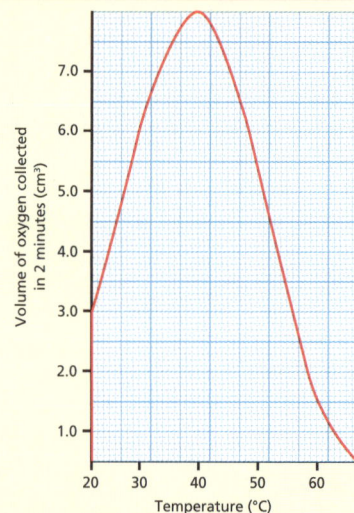

Common errors

- The hydrogen peroxide solution must be left in the water bath to reach the correct temperature before the potato is added.
- The delivery tube should be placed under the measuring cylinder immediately after the potato cylinders are added and the cork inserted into the boiling tube. If it is already under the cylinder, air will be pushed into the cylinder.

Investigating other variables

This apparatus can be used to investigate the effect of other variables on the rate of reaction catalysed by the enzyme catalase. These include the following:

- The **effect of pH,** by adding buffer solutions of various pH values to the hydrogen peroxide solution. The temperature is kept constant at the optimum value.
- **Substrate concentration** can be varied by using different concentrations of hydrogen peroxide solution.
- **Enzyme concentration** can be varied by using different numbers of potato cylinders.

Required practical 2: Root tip squash

Aim

To prepare a stained squash of cells from plant root tips to identify the stages of mitosis and calculate a mitotic index

My lab notes

Introduction

Onion or garlic bulbs will produce roots if suspended with their bases in water for about one week. These roots will contain meristematic tissue near the tip that can be used to view cells in stages of mitosis. The root tips need to be stained and squashed so that they can be viewed.

Example method

1. Cut about the last 2 cm of a root and leave the cutting in hydrochloric acid for 15 minutes.
2. Rinse the root tip in distilled water on a watch glass.
3. Cut off the last 1–2 mm of the root tip and place this on a microscope slide.
4. Cover the section with a dye, such as toluidene blue stain, for several minutes.
5. Break up the tip with the mounted needle to separate the cells.
6. Add a cover slip, and with gentle finger pressure 'spread' the material and blot at the same time by using a folded filter paper between finger and slide.
7. Use the high-power objective lens of a microscope to count the cells that are in different stages of mitosis.
8. Repeat for several fields of view and record the data in a suitable table.

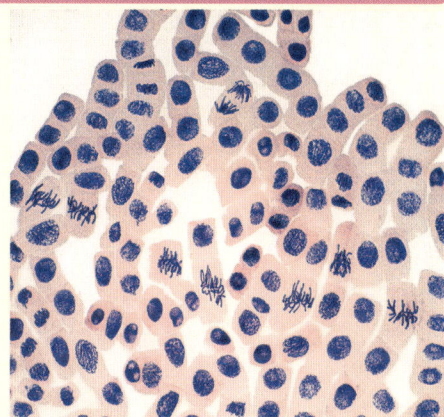

Analysis

The results table shows the number of cells in each stage of the cell cycle, as seen on a root tip squash.

Calculate the mitotic index using:

Mitotic index = number of cells in stages of mitosis ÷ total number of cells

So, in this example it is 50 ÷ 145 = 0.34

Stage	Number of cells			
	Field 1	Field 2	Field 3	Total
Interphase	32	27	36	95
Prophase	5	7	6	18
Metaphase	2	2	0	4
Anaphase	4	2	3	9
Telophase	6	8	5	19

The results also give an estimate of the relative length of time that cells spend in each stage of the cell cycle. For example, cells will spend 95 ÷ 145 × 100 = 66% of the time in interphase.

Common errors

- The cells are overlapping each other, making it impossible to see if the chromosomes are visible – this is because the tissue has not been teased apart enough or not squashed firmly enough.
- None of the cells appear to be in mitosis and there is also xylem tissue visible – this is because the wrong region of the root is being viewed. The area is too far from the root tip, which could be because the wrong end of the cutting has been selected in step 3 of the method.
- Counting errors – it may be easier to take a photograph of the cells and use this to count the cells in each stage.

Further investigations

The mitotic index is an indication of how fast cells are dividing. For example, cancer cells often have a high mitotic index. This method can be used to compare the mitotic index in different aged roots, from garlic or onion, to see how the rate of growth changes. It can also be used to compare growth rates between different roots from different varieties or species.

Required practical 3: Water potential of potato tissue

Aim

To produce a dilution series of a solute to construct a calibration curve and use this to identify the water potential of plant tissue

My lab notes

Example method

1. Use 1.0 mol dm^{-3} sucrose solution and water to make up 20 cm^3 of sucrose solution of different concentrations using the proportions shown in the table.

Concentration of sucrose solution (mol dm^{-3})	0	0.2	0.4	0.6	0.8
Volume of 1.0 mol dm^{-3} sucrose solution used (cm^3)	0	4	8	12	16
Volume of water used (cm^3)	20	16	12	8	4

2. Pour each of the solutions into separate boiling tubes and keep in a water bath set at 30°C for 5 minutes.
3. Using a cork borer, cut six cylinders from a potato tuber and cut the cylinders to the same length.
4. Blot the cylinders dry with a paper towel, i.e. roll each chip until it no longer wets the paper towel and then measure the mass of each potato cylinder.
5. Transfer each potato cylinder to its boiling tube and leave for 30 minutes.
6. Remove the cylinders from the boiling tubes, blot them dry as before and then reweigh them.

Analysis

In a results table, calculate the percentage change in mass for each potato cylinder.

Then plot a graph of the percentage change in mass against the concentration of the sucrose solution.

Concentration of sucrose solution (mol dm^{-3})	Mass of potato cylinder (g)			
	Start	Final	Change in mass	Percentage change in mass
0.0	1.30	1.51	0.21	16.2
0.2	1.35	1.50	0.15	11.1
0.4	1.30	1.35	0.05	3.8
0.6	1.34	1.27	−0.07	−5.2
0.8	1.30	1.18	−0.12	−9.2

Where the line crosses the x-axis, there is no net loss or gain of water from the potato cylinders. This means the water potential of the potato tissue is equivalent to the concentration of the sucrose solution at that point. On the graph this is 0.45 mol dm^{-3}.

Looking in a data book, this value is equivalent to a water potential of −1.24 KPa.

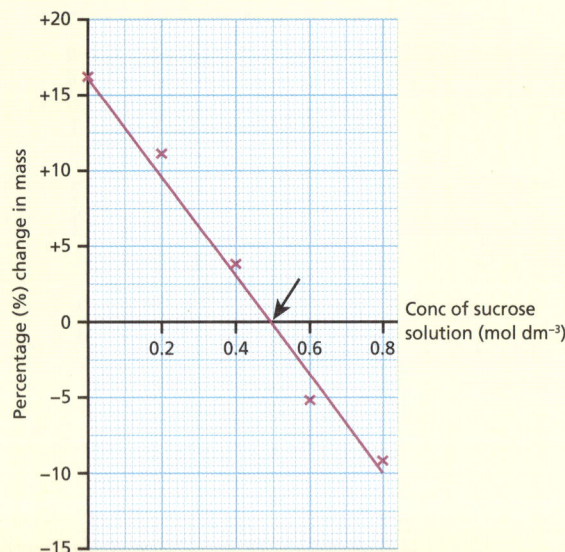

Further investigations

This method can be used to compare the water potential in potato tubers stored for different lengths of time – over time the starch in the tubers is often converted to sugars, altering the water potential. It can also be used to compare the water potential of different tissues such as sweet potato, carrots and parsnips.

A modification of this method can be used, which involves seeing if a drop of the original sucrose solution will fall or rise in the solution that has been incubated with the potato chip – if the drop of the original solution rises, the incubated solution must have become denser due to the loss of water to the potato by osmosis.

Required practical 4: Membrane permeability

Aim

To investigate the effect of a named variable on the permeability of cell-surface membranes

My lab notes

Introduction

Beetroot tissue is a good choice of material to investigate membrane permeability. The cell vacuoles contain high concentrations of a purple pigment called betalain. This pigment cannot move across undamaged plasma membranes. The permeability of the plasma membranes can be assessed by measuring the amount of betalain that leaks into a bathing solution. The colour can be measured quantitatively by using a colorimeter. In this experiment, the effect of temperature is investigated.

Example method

1. Cut five 1 cm cylinders of beetroot tissue using a cork borer.
2. Place all the cylinders in a beaker of distilled water and leave for some time to remove any betalain released when the beetroot was cut.
3. Wash and blot dry (with filter paper or a tissue) the five pieces of beetroot.
4. Fill five boiling tubes each with 10 cm³ of distilled water and place them into five water baths set at different temperatures (e.g. 0°C, 20°C, 40°C, 60°C, 80°C) for 5 minutes.
5. Add a cylinder of beetroot to each boiling tube and leave for 30 minutes.
6. Remove the beetroot pieces and then shake the tubes to disperse the betalain.
7. Set a colorimeter to percentage absorbance on the green filter and calibrate using a cuvette with distilled water.
8. Pour each of the solutions from the boiling tubes into cuvettes and read the percentage absorbance.

Independent variable	Temperature
Dependent variable	Absorbance of the bathing solution
Control variables	Size of beetroot cylinders
	Volume of bathing solution

Analysis

Plot a graph of the percentage absorbance of the solution against temperature. As the permeability of the membrane increases, more betalain diffuses out of the cylinders into the bathing solution. This produces a darker colour and a higher reading for absorbance on the colorimeter.

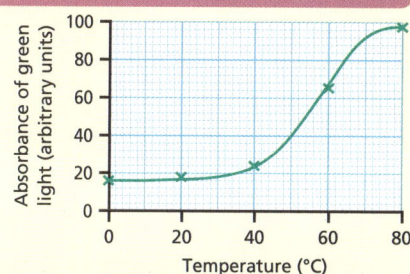

Common errors

- Leaving the beetroot skin on the ends of some of the cylinders – this will reduce the loss of pigment.
- Failure to wash the cylinders after cutting – this would introduce betalain into the bathing solution from the cut cells.
- Cutting cylinders from different parts of the beetroot tuber – different parts will have different properties.

Further investigations

- The effect of other factors on the permeability of the membrane can be investigated, such as ethanol or detergents. Both of these can disrupt the cell membrane.
- Calcium can give cell membranes some protection from thermal damage – this could be investigated.

Required practical 5: Organ dissection

Aim

To dissect an animal or plant gas exchange system or mass transport system, or of an organ within such a system

My lab notes

Introduction

A sheep's heart is a good organ to dissect – they are readily available and large enough to see the main structures. It is important to examine and identify the main external structures before any dissection takes place.

Example method

1. Place the heart on the dissecting board or tray.
2. Examine the outside of the heart, making sure that the ventral (front) side is uppermost. This will make the arteries at the top of the heart visible.
3. Identify the coronary arteries running over the surface of the ventricles and the position of the septum, which runs at an angle downwards to separate the right and left sides.
4. Squeeze both the right and the left sides of the heart to compare the thickness of the two ventricles.
5. Locate the right and left atria on top of the ventricles and note how thin they are compared to the ventricles.
6. Find the veins that are attached to each atrium.
7. Identify the arteries leaving the ventricles. The best way to do this is to insert a blunt needle into the artery and push it down into the ventricle. If it finishes on the right side, it is in the pulmonary artery; if it is on the left side, it is in the aorta. At this stage, it is good practice to make a biological drawing of the ventral view of the heart. The next steps involve opening the heart:
8. On the left side of the heart, cut upwards from the base of the ventricle, just to the side of the septum. Continue the cut into the left atrium to open that up. Open out the heart to expose the inside of the left ventricle and atrium. The bicuspid valve should be visible.
9. On the right side of the heart, make a similar cut upwards from the base of the ventricle, just to the side of the septum. This time cut up into the pulmonary artery rather than the atrium. Open up the heart to see the entrance to the right ventricle and the semilunar valve in the pulmonary artery.
Now make a biological drawing of the opened heart.

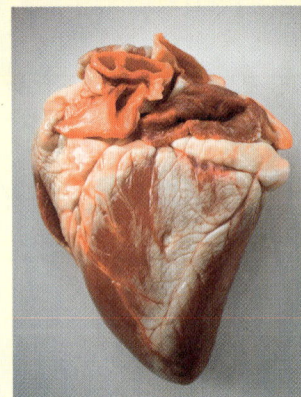

Drawing of the dissection of the sheep's heart viewed from the ventral side.

Magnification × 0.5

Pulmonary artery — Aorta
Semilunar valve — Pulmonary vein
Vena cava
Right atrium — Left atrium
Tricuspid valve — Bicuspid valve
Cordae tendoneae — Wall of left ventricle
Branch of coronary artery — Papillary muscle
Wall of right ventricle — Septum

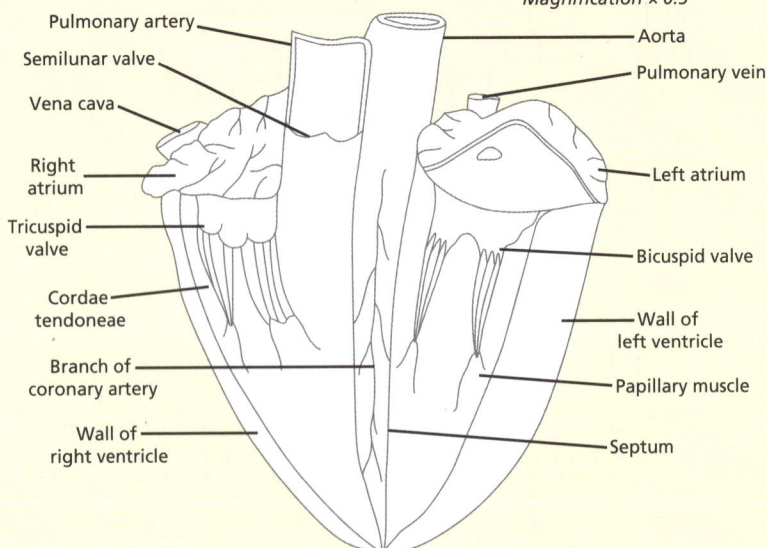

All biological drawings should:
- be drawn with thin, continuous pencil lines with no sketching
- have no shading, but cross-hatching is acceptable
- include labels with non-overlapping label lines
- have a title and include a magnification.

Required practical 6: Testing antimicrobial substances

Aim

To use aseptic techniques to investigate the effect of antimicrobial substances on microbial growth

My lab notes

Introduction

It is possible to use cultures of microorganisms such as bacteria to investigate the effects of antimicrobial substances. It is important to use pure cultures of non-pathogenic bacteria in these studies.

Example method

1. Set up a sterile area by wiping all work surfaces with disinfectant.
2. Transfer some of a bacterial culture onto an agar plate using a sterile pipette. Spread it out using a sterile spreader.
3. Obtain paper discs that have been pre-treated with different antibiotics.
4. Space the discs apart on the agar plate using sterile forceps.
5. Lightly tape a lid onto the Petri dish, invert and incubate at 25°C for 24 to 48 hours.

Analysis

The antibiotics will diffuse into the agar and may kill the bacteria, producing a clear zone. The larger the clear zone, the more effective the antibiotic is at inhibiting bacterial growth. If no clear zone is present, it means that the bacteria are resistant to that type of antibiotic. The sizes of the clear zones produced by the different types of antibiotic can be compared.

To obtain a quantitative measure of the ability of each antibiotic to kill the bacteria:

- measure the diameter of each clear zone with a ruler and use this to calculate the area of the clear zone using πr^2.
- plot a bar chart to display the areas of the clear zones.

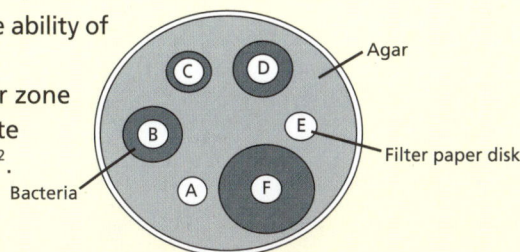

Precautions

When working with microorganisms, it is important to follow aseptic techniques. This will avoid any possible contamination of samples and risks to health.

Aseptic techniques include:

- washing hands thoroughly to disinfect them before and after the investigation
- disinfecting work surfaces to kill any pathogenic bacteria
- fully sterilising all apparatus before use, including the spreaders used to transfer bacteria and the forceps used to move the antibiotic discs
- lifting the lids of the Petri dishes for the minimum amount of time to insert the bacteria and discs
- lightly taping the lids onto the Petri dishes, rather than sealing them; this ensures that oxygen is available to the bacteria and means that harmful anaerobic bacteria are less likely to grow
- incubating the dishes at 25°C rather than at body temperature, so that human pathogens are less likely to grow
- making sure that all agar plates are destroyed after use.

Further investigations

Using agar plates inoculated with bacterial cultures, it is possible to investigate a range of different antimicrobial substances:

- A range of commercially available disinfectants of different costs can be compared to see how effectively they kill a strain of bacteria.
- The technique of serial dilution can be used to investigate the effect of different concentrations of a disinfectant on the ability to kill bacteria.
- Different toothpastes can be compared to see if they kill bacteria.
- Various plant extracts, such as garlic, mint, basil and mustard, are thought to have antimicrobial properties. A range of extracts could be tested to see if they kill bacteria.

Required practical 7: Chromatography

Aim

To use chromatography to investigate the pigments isolated from the leaves of different plants

My lab notes

Introduction

Chromatography can be used to separate and identify the different pigments that are found in the chloroplasts of leaves. In this experiment, two different species of leaves are used to compare the pigments present.

Example method

1. Put 3 cm^3 of solvent into two boiling tubes. Put a bung in the top of each tube and label the tubes A and B.
2. Cut two pieces of chromatography paper to fit into the boiling tubes. Rule pencil lines 2 cm from the bottom of each filter paper.
3. Grind a few leaves from species A in a mortar with 10 cm^3 of propanone to extract the pigments.
4. Use a fine pipette tip or capillary tube to take up a small amount of the extract. Place one small drop of this extract in the centre of the pencil line and allow to dry before adding another drop on top. Build up a pigment spot that is as small as possible but concentrated.
5. Carefully pour the chromatography solvent into the boiling tube to a depth of about 1 cm.
6. Suspend the chromatography paper inside the boiling tube. The bottom of the paper should be dipped into the solvent, but the pigment spot must not be immersed in the solvent. Insert the bung.
7. The solvent front should rise up the paper, and when it is close to the top of the tube, remove it from the solvent and quickly mark the position of the front using a pencil.
8. Repeat steps 3–7 with leaves from species B.

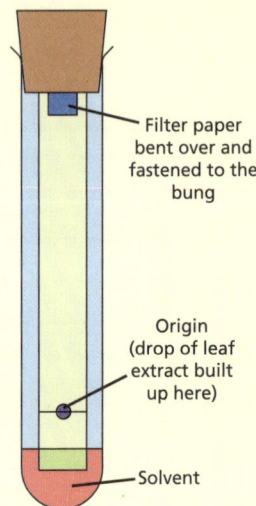

Filter paper bent over and fastened to the bung

Origin (drop of leaf extract built up here)

Solvent

Analysis

The different coloured pigments from the leaf travel at different rates through the chromatography paper and so separate out.

To identify the pigments, calculate an R_f ratio for each pigment. This is a ratio of the distance travelled by a pigment compared to the solvent front:

$$R_f = \frac{a}{b}$$

The pigments can then be identified by using R_f values from a table.

Solvent front

Position of pigment

Origin

Pigment	Colour of spot	R_f value
Carotene	Yellow-orange	0.95
Phaeophytin	Grey-yellow	0.83
Xanthophyll	Yellow-brown	0.71
Chlorophyll a	Blue-green	0.65
Chlorophyll b	Light green	0.45

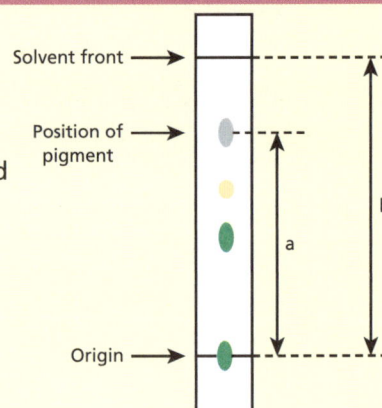

Further investigations

- Leaves from different regions of the same plant can be tested.
- Pigments in different coloured leaves can be identified.
- Thin-layer chromatography (TLC) can be used rather than paper chromatography. TLC uses a thin layer of an inert substance (e.g. silica) supported on a glass plate. It often produces better separation of pigments than paper chromatography.

Required practical 8: Dehydrogenase activity in chloroplasts

Aim

To investigate the effect of a named factor on the rate of dehydrogenase activity in extracts of chloroplasts

My lab notes

Introduction

Dehydrogenase enzymes are found in plant chloroplasts and are involved in the light-dependent stage of photosynthesis. Dehydrogenase enzymes catalyse the uptake of electrons by NADP.

A redox indicator, such as DCPIP, can be used to accept the electrons instead of NADP. This turns DCPIP from blue to colourless when it is reduced. This method can be used to investigate the effect of light intensity on the rate of dehydrogenase activity.

Example method

1. Remove stalks from spinach leaves and grind the leaves using a pestle and mortar with some chilled isolation solution.
2. Filter the sample through a muslin cloth and funnel into a beaker. Keep the beaker chilled in an ice water bath. This is the chloroplast suspension.
3. Add 5 cm^3 of water to 1 cm^3 of chloroplast suspension in a test tube. This is the standard tube.
4. Add 5 cm^3 of DCPIP solution and 1 cm^3 of chloroplast suspension to another test tube and position a lamp 10 cm away from the test tube.
5. Record how long it takes for the colour of the tube to change from blue-green to green. Use the standard tube to judge the final colour.

Independent variable	Light intensity
Dependent variable	Time taken for DCPIP to be decolourised
Control variables	Concentration and volume of chloroplast suspension

6. Repeat steps 4 and 5 for different distances from the lamp up to 100 cm.

Analysis

- To convert the distance into a light intensity, use the inverse-squared law, i.e. the light intensity is proportional to $\frac{1}{distance^2}$
- Calculate the rate of reaction using $\frac{1}{time\ taken}$

Distance from lamp (m)	Light intensity $(\frac{1}{d^2})$	Time taken to change colour (s)	Rate of reaction $(\frac{1}{t})$
0.10	100	25	0.04
0.25	16	27	0.037
0.50	4.0	40	0.025
0.75	1.8	55	0.018
1.00	1.0	70	0.014

Plot a graph of rate of reaction against light intensity. It should show that at low light intensities, this is the limiting factor. However, at higher light intensities some other factor becomes limiting (possibly the number of chlorophyll molecules to trap light).

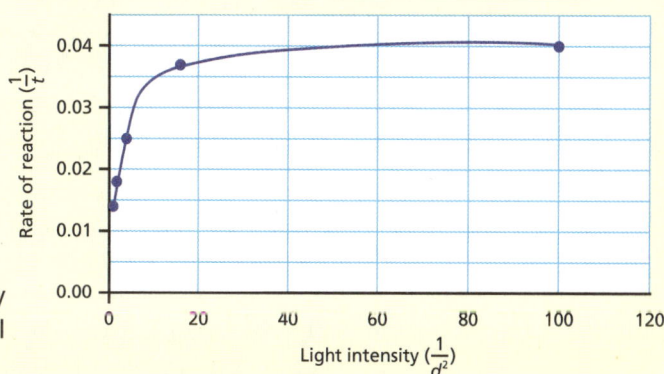

Further investigations

- This method can be adapted to investigate the effect of other variables on the rate of dehydrogenase activity, such as temperature.
- Certain inhibitors, such as ammonium hydroxide, can inhibit the action of dehydrogenase enzymes and so this effect can be studied.
- A number of weedkillers act by blocking the electron transport chain. This method can be used to identify whether a weedkiller works in this way as it will prevent the reduction of DCPIP.

Required practical 9: Respiration in yeast

Aim

To investigate the effect of a named variable on the rate of respiration of cultures of single-celled organisms

My lab notes

Introduction

The single-celled organism chosen for this investigation is yeast, which can respire aerobically and anaerobically. During aerobic respiration, electrons pass through the electron transport chain to synthesise ATP. In this investigation, the electrons are accepted by a substance called methylene blue, which changes colour from blue to colourless. This experiment investigates the effect of different substrates on the rate of respiration of yeast.

Example method

1. Label four test tubes with the names of four different sugars: glucose, sucrose, maltose and fructose.
2. Add 2 cm^3 of suspension yeast and 3 cm^3 of the appropriate sugar solution to each tube.
3. Place all four tubes in a water bath at 35°C for several minutes.
4. Add 2 cm^3 methylene blue to the test tube containing glucose.
5. Immediately shake this tube for 10 seconds and replace the tube in the water bath. Note the time and do not shake this tube again.
6. Record how long it takes for the blue colour to disappear in the tube.
7. Repeat steps 4 to 6 for the tubes containing the other sugars.

Independent variable	Type of substrate
Dependent variable	Time taken for methylene blue to decolourise
Control variables	Concentration and volume of substrate solution and methylene blue solution
	Temperature

Analysis

Calculate the rate of reaction by using $\frac{1}{time\ taken}$ and plot a bar chart of rate of reaction against type of substrate.

Substrate	Time taken to decolourise (s)	Rate of reaction ($\frac{1}{t}$)
glucose	22	0.045
sucrose	60	0.017
maltose	35	0.029
fructose	68	0.015

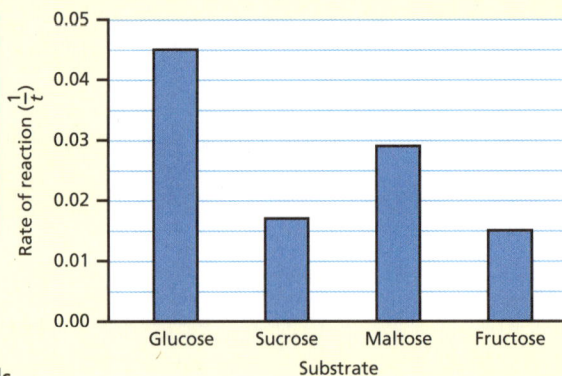

The results of the investigation depend on two factors:
* How rapidly the substrate molecules can enter the yeast cells.
* How readily each substrate can enter the respiratory pathway.

Possible improvements

The main difficulty with this investigation is judging the endpoint. Two improvements could assist with this:
* Prepare a standard tube containing yeast and sugar solution, but no methylene blue, to use as a comparison to judge the endpoint.
* A colorimeter can be used to judge the colour change but this is difficult because yeast makes the solution very cloudy. Trapping the yeast in immobilised beads can get round this problem.

Further investigations

* This method can be adapted to investigate the effect of other variables on the rate of respiration of yeast, such as temperature, pH and substrate concentration.
* Salt is often used to slow the respiration of yeast during bread making. The effect of salt on the respiration rate can be investigated.
* Various inhibitors of respiration, such as heavy metal salts, can be tested on yeast.

150

Required practical 10: Kinesis in woodlice

Aim

To investigate the effect of an environmental variable on the movement of an animal using either a choice chamber or a maze

My lab notes

Introduction

Woodlice are good organisms to use when studying the effect of the environment on their behaviour. When they are in an adverse environment, they decrease the rate at which they change direction and they move faster. This means they spend more time in a favourable environment. This response is called **kinesis**. A choice chamber can be used to investigate this response.

Example method

1. Set up the choice chamber with nothing in the base.
2. Place 20 woodlice in the chamber through the central hole.
3. Wait 10 minutes and then record the number of woodlice in the left and right halves of the choice chamber.
4. Set up the choice chamber to have four sections as follows: dark and dry; dark and humid; light and dry; light and humid. Use dark paper to produce dark conditions, use water underneath to make humid areas, and use a drying agent underneath (such as anhydrous calcium chloride) to make dry areas.
5. Place 20 woodlice in the centre of the choice chamber and leave for 10 minutes.
6. Record how many woodlice are in each quarter.

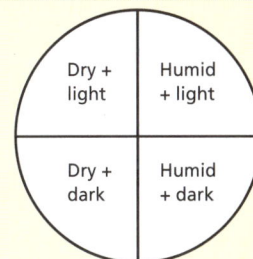

Dry + light	Humid + light
Dry + dark	Humid + dark

Analysis

The first three steps in the method are to see if the woodlice distribute at random with the same conditions in each section. If they do, analyse the results after step 6 of the method.

Plot a bar chart to show the number of woodlice in each quarter.

To test to see if the woodlice are distributed at random in the four sections, perform a Chi-squared test using the results.

Conditions	Observed number (O)	Expected number (E)	$O - E$	$(O - E)^2$	$\frac{(O - E)^2}{E}$
Dry and light	1	5	−4	16	3.2
Dry and dark	3	5	−2	4	0.8
Humid and light	3	5	−2	4	0.8
Humid and dark	13	5	8	64	12.8

The Chi-squared formula is: $\chi^2 - \Sigma \frac{(O - E)^2}{E}$

So, Chi-squared value = 3.2 + 0.8 + 0.8 + 12.8 = 17.6

At three degrees of freedom, the critical value at $p = 0.05$ is 7.8

As 17.6 is greater than 7.8, the null hypothesis is rejected. The observed numbers are significantly different to the expected numbers, i.e. the woodlice are not distributed at random.

Further investigations

- Using the choice chamber and one woodlouse, it is possible to record the movements of the woodlouse. This can be done by tracing the movements of the woodlouse by marking the lid of the choice chamber with a temporary marker pen. This can be used to compare how often the woodlouse turns when in favourable or unfavourable conditions.
- The movement of a woodlouse in the choice chamber can be recorded by making a video using a smartphone. This can then be used to calculate the speed of the woodlouse in favourable and unfavourable conditions.

Required practical 11: Glucose calibration curve

Aim

To produce a dilution series of a glucose solution and use colorimetric techniques to produce a calibration curve to identify the concentration of glucose in an unknown 'urine' sample

Introduction

Simply looking for a red colour in a Benedict's test is a qualitative approach. A quantitative Benedict's test can be performed by creating a calibration curve, using known concentrations of glucose solution. The extent of the colour change to red depends on the concentration of glucose in the sample, so the concentration can be found by measuring the absorbance in a colorimeter. Once a calibration curve is constructed using known concentrations of glucose, it can be used to find the unknown concentrations of solutions, such as a 'mock' urine sample.

Example method

Preparing the glucose calibration curve

1. Dilute a glucose standard (10 mmol dm^{-3}) with water using the volumes in the table to produce a range of glucose concentrations. Put each concentration in a separate test tube.

Concentration of glucose solution produced (mmol dm^{-3})	0.0	2.0	4.0	6.0	8.0	10.0
Volume of water used (cm^3)	2.0	1.6	1.2	0.8	0.4	0.0
Volume of glucose standard used (cm^3)	0.0	0.4	0.8	1.2	1.6	2.0

2. Add 2 cm^3 of Benedict's solution to each tube and place all the test tubes into a water bath at 90°C for four minutes.
3. Allow the tubes to cool before taking readings from the colorimeter.
4. Use the contents of the 0.0 mmol dm^{-3} glucose solution tube as a blank to calibrate the colorimeter to zero absorbance.
5. Place the remaining samples in cuvettes into the colorimeter and read their absorbances.

Testing the 'urine' samples

Repeat steps 2 to 5 using 2 cm^3 of a mock urine sample.

Analysis

Use the results from the standard glucose solutions to plot a graph of absorbance against concentration of glucose. This is a calibration curve.

Use the calibration curve to read off the glucose concentration from the percentage absorption of the mock urine sample. In this case, the concentration would be 5 mmol dm^{-3}.

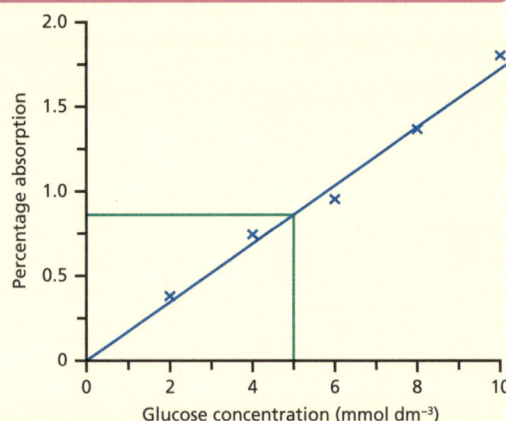

Further investigations

- The same technique can be used to test a 'mock' urine sample from a person suspected of having diabetes.
- Calibration curves can be used to compare the sweetness of different fruits, vegetables, fruit juices or wines.
- The ripening process of fruits can be studied by measuring the change in sugar content of the fruits over time. The effect of keeping the fruits with a ripened fruit that is giving off the plant hormone ethene could be investigated.

Required practical 12: Light intensity and leaf area

Aim

To investigate the effect of a named environmental factor on the distribution of a given species

My lab notes

Introduction

Leaves that grow in shaded areas ('shade leaves') often have different characteristics to leaves that grow in full sunlight ('sun leaves'). There are often differences in thickness, with sun leaves being thicker than shade leaves. Shade leaves may also have a larger area for absorbing light energy for photosynthesis. Also, smaller sun leaves will provide less surface area for the loss of water through transpiration. The relationship between light intensity and leaf area can be investigated in this practical.

Example method

1. Choose a species of plant that grows in an area from shade out into full sunlight.
2. Run a transect line in the air from the shade into the full sunlight.
3. Identify plants along the transect line that are close to the height of the transect line.
4. For each plant, measure the width and length of a corresponding leaf (e.g. the fourth leaf down from the tip of the shoot).
5. At the site of each plant, take a light meter reading, pointing the light meter upwards and making sure that you do not shade it.

Analysis

Calculate an estimate of the leaf area of each leaf by using this formula:

leaf area = length × width × 0.75

Then plot a scatter graph of leaf area against light intensity reading.

To see if there is a statistically significant negative correlation between leaf area and light intensity, perform a Spearman's correlation coefficient calculation.

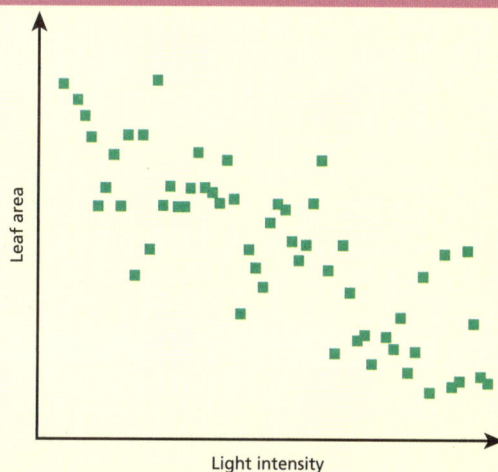

Light intensity	Rank	Leaf area	Rank	d	d^2
700	10	127	1	9	81
850	9	94	3	6	36
950	8	103	2	6	36
1050	7	51	10	−3	9
1100	6	65	5	1	1
1150	5	64	6	−1	1
1250	4	76	4	0	0
1394	3	53	9	−6	36
1450	2	63	7	−5	25
1500	1	52	8	−7	49

The formula for Spearman Rank is: $\rho = 1 - \dfrac{6\sum d^2}{n(n^2-1)}$ So, $\rho = 1 - \dfrac{6 \times 274}{10 \times 99} = 0.66$

Looking at the coefficient tables with 10 degrees of freedom, it shows that 0.66 lies between $\rho = 0.05$ and 0.01, so there is a significant negative correlation between leaf area and light intensity.

Points to note

- With all ecological investigations, it is difficult to control all variables. It is important to take all readings in the same time period, on a clear day, to avoid changes in sunlight. However, variations in soil composition are difficult to rule out.
- Whilst the Spearman rank test indicates a correlation between the two factors, it does not prove that the leaf area is directly affected by light intensity.

THIS PAGE HAS DELIBERATELY BEEN LEFT BLANK

Collins

A-level
Biology
Practice paper for AQA

Paper 1　　　　　　　　　　　　Time allowed: 2 hours

Materials

For this paper you must have:
- a ruler with millimetre measurements
- a scientific calculator.

Instructions
- Use black ink or black ball-point pen.
- Fill in the box at the bottom of this page.
- Answer **all** questions.
- Show all your working.

Information
- The marks for the questions are shown in brackets.
- The maximum mark for this paper is 91.

Name: ....

Answer **all** questions in the spaces provided.

0 1   **Figure 1** shows the structure of a fibrous protein called collagen.

Figure 1

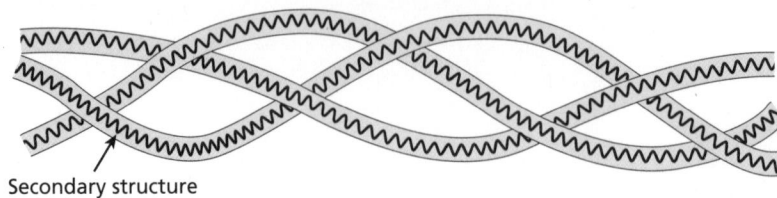

Secondary structure

The secondary structure of the protein is shown in the fibres.

0 1 . 1 State the name of this type of secondary structure.

[1 mark]

Collagen is rich in the amino acids proline and glycine. The structure of these molecules is shown in **Figure 2**. They can join together to form a peptide bond.

Figure 2

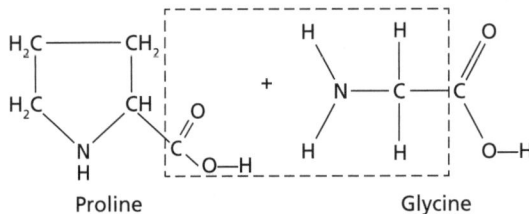

Proline   Glycine

0 1 . 2 Which smaller molecule is eliminated in this reaction?

[1 mark]

0 1 . 3 Draw a ring round the R group of the glycine molecule.

[1 mark]

$\boxed{0\ 1}\cdot\boxed{4}$ In the space below, draw the chemical bond formed when these two amino acids are joined together. Only draw the portions of the molecule **within** the box in **Figure 2**.

[2 marks]

$\boxed{0\ 1}\cdot\boxed{5}$ Collagen is the main protein found in tendons. Tendons connect muscles to bone. When muscles contract, tendons will transfer the force to bones to enable movement.

Explain how collagen's structure is adapted for this function.

Use information from **Figure 1** in your answer.

[2 marks]

0 2 Albumen is a **globular** protein found in egg white. When heated, a permanent change occurs to the structure of the albumen molecule.

This causes the egg white to change from a transparent liquid to an opaque white solid.

0 2 . 1 What name is given to the change in protein structure?

[1 mark]

....

0 2 . 2 Explain how a raised temperature can change the tertiary structure of albumen and cause the change in egg white.

[3 marks]

....

....

....

....

0 2 . 3 Describe how you would carry out a biochemical test to determine that albumen was indeed a protein.

[2 marks]

....

....

....

0 2 . 4 During digestion, polypeptides are hydrolysed to smaller molecules that can be absorbed across cell membranes.

Compare and contrast how endopeptidases and exopeptidases break down polypeptide molecules.

[3 marks]

....

....

....

....

Invertase is a yeast-derived enzyme that hydrolyses **sucrose** into **glucose** and **fructose**.

Figure 3 shows the effect of the concentration of sucrose on the rate of reaction of invertase.

Figure 3

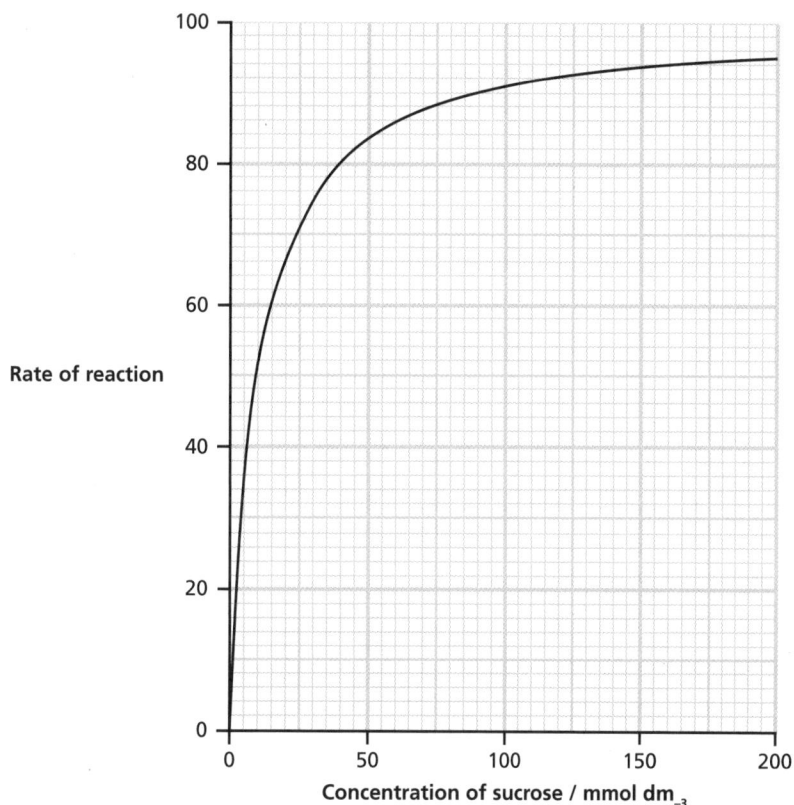

Rate of reaction (y-axis)

Concentration of sucrose / mmol dm$_{-3}$ (x-axis)

0 3 · 1 Calculate the percentage increase in the rate of reaction between 25mmol dm$_{-3}$ and 50 mmol dm$_{-3}$

[2 marks]

.... %

0 3 · 2 For each of the control variables below, suggest **how** they could be controlled.

[1 mark]

Temperature ....

[1 mark]

pH ....

0 3 · 3 Explain why the rate of reaction levels off at high concentrations of sucrose.

[2 marks]

....

....

0 3 . 4 The yeast from which invertase is extracted is a strain called *Saccharomyces cerevisiae*. Complete **Table 1** to show its classification.

[2 marks]

Table 1

Taxon	Name of Taxon
	Eukaryota
Kingdom	Fungi
	Ascomycota
Class	Saccharomycetes
	Saccharomycetales
Family	Saccharomycetaceae
Genus	
Species	Cerevisiae

0 4   **Figure 4** shows a transmission electron micrograph of a macrophage cell. The micrograph has been magnified **6000** times.

Figure 4

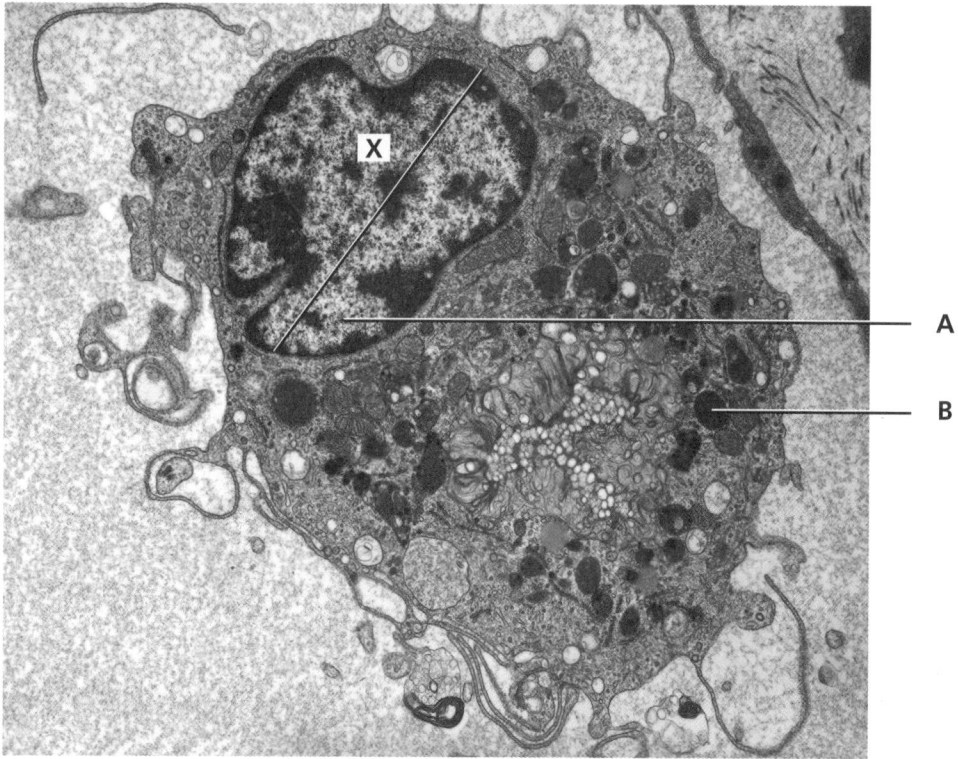

0 4 · 1 Give the name and function of structure **A**.

[1 mark]

Name of **A** ....

Function of **A** ....

0 4 · 2 Structure **B** is a lysosome. Explain why this cell has a large number of these organelles.

[2 marks]

....

....

....

0 4 · 3 Measure the diameter of the image of structure **A** along the line indicated by an **X** in **Figure 4**.

Calculate the **actual** diameter of structure **A in** μm using the information given.

[2 marks]

Actual diameter of **X** ....

0 5 . 1 Describe how you would prepare a specimen of stained squashes of cells from plant root tips such as garlic or onion for observation under a light microscope. You should name a suitable stain in your answer.

[5 marks]

[Extra space]

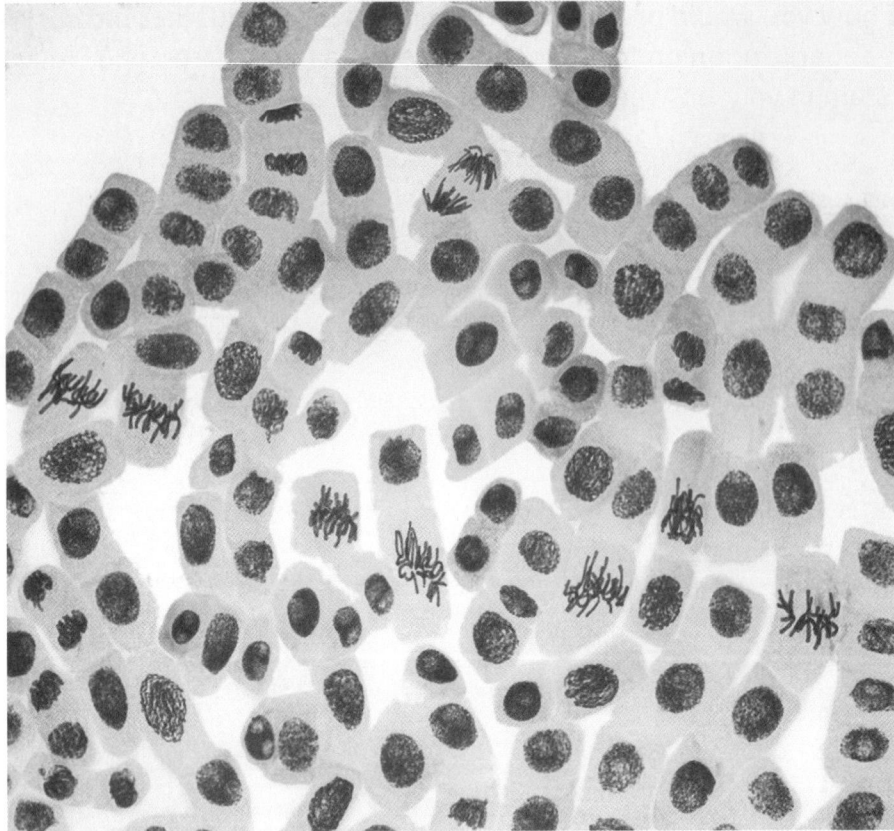

Assuming there are 130 cells in this field of view, calculate the **mitotic index** of the specimen.

[2 marks]

Answer .... %

0 6   A student used this method to investigate **plasmolysis** in plant cells.

- A strip of epidermis from the inner surface of a red onion's fleshy storage leaves was peeled away.

- Using forceps, a small piece of epidermis was transferred to a microscope slide and three drops of distilled water added.

- After placing a coverslip on top, the cells were examined under a microscope.

Figure 6 shows the appearance of the cells.

Figure 6

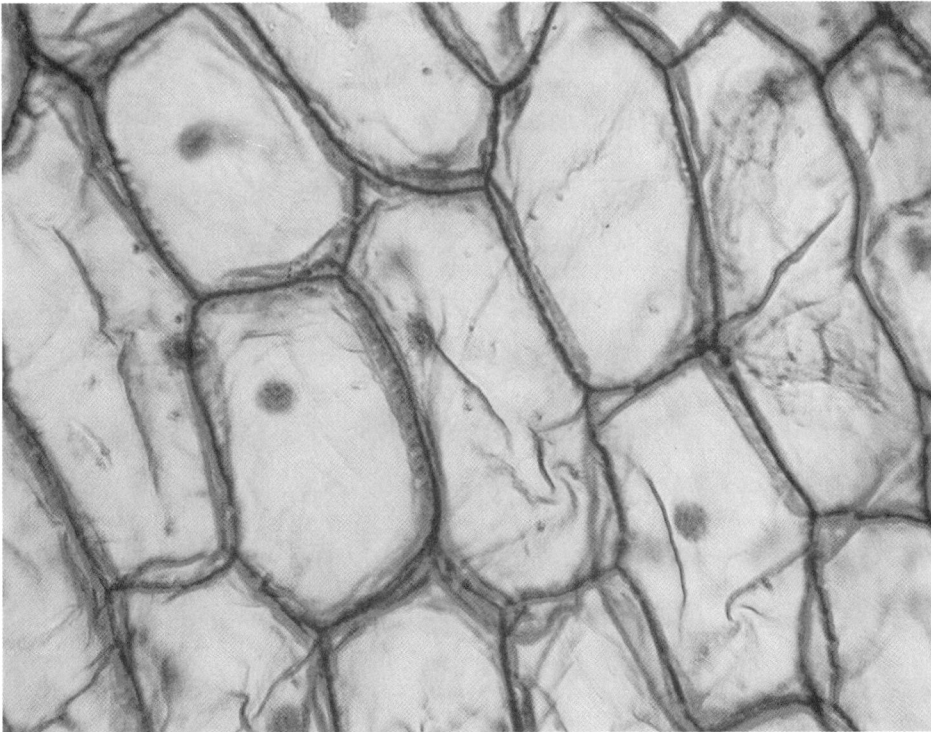

The procedure was repeated, but this time the epidermis was mounted in 1 M sucrose solution. Many of the cells had a different appearance. They had been **plasmolysed**.

0 6 · 1 In the space below, draw the likely appearance of a single cell placed in 1 M sucrose solution. Label the cytoplasm.

[2 marks]

0 6 · 2 Plasmolysis rarely occurs in nature. However, plants that live in salt marshes have to resist plasmolysis. Explain, in terms of water potential, why this is so.

[2 marks]

0 6 · 3 Suggest a reason for using red onion rather than white onion for the experiment.

[1 mark]

The student counted the number of cells they could see in the microscope's field of view. They also counted the total number of cells.

They then set up different slides where onion cells were mounted in a range of different concentrations of sucrose solution. Their results are shown in **Table 2**.

Table 2

Concentration of sucrose solution/M	Number of plasmolysed cells	Total number of cells	% plasmolysed cells
0.05	0	40	0.0
0.15	2	57	3.5
0.25	5	48	
0.35	32	50	64.0
0.45	50	64	78.1
0.55	50	50	100

0 6 . 4 Calculate the percentage of plasmolysed onion cells in 0.25 M sucrose solution.

[1 mark]

.... %

0 6 · 5 Use **Figure 7** to draw a graph to display these results. The axes have been labelled for you. Include a line of best fit.

[3 marks]

Figure 7

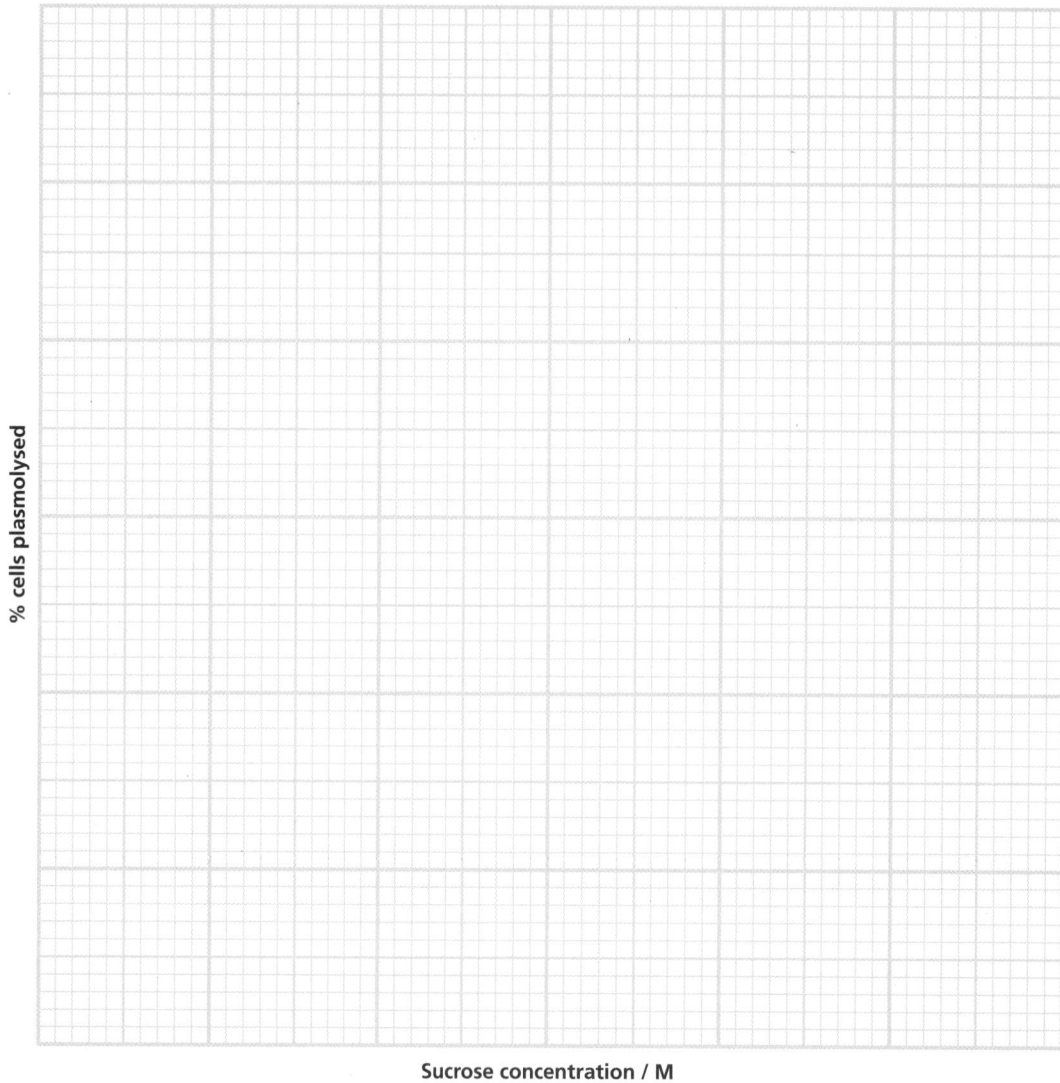

When the contents of a plant cell are isotonic with the external solution, a state of **incipient plasmolysis** is the result. This is the concentration where cytoplasm is just held in place against the cell wall. Experimentally, this can be estimated by finding the external concentration that results in 50% of cells being plasmolysed. Thus, you can find the concentration of the cell's sap.

0 6 · 6 Draw two lines on your graph to show the data values at incipient plasmolysis. Use your graph to estimate the concentration of cell sap in red onion leaf cells.

[2 marks]

Figure 8 shows how the metabolic rates of various mammals change with their body mass.

Figure 8

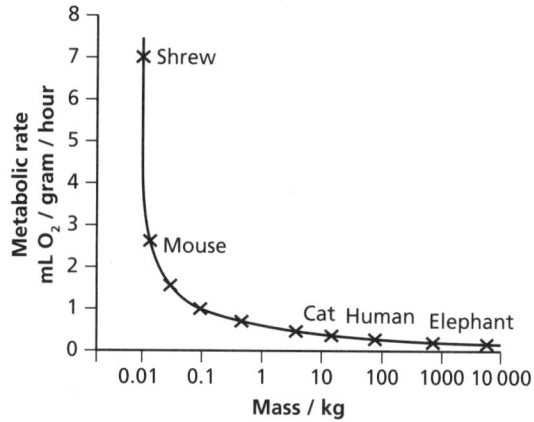

The body mass is plotted on a logarithmic scale. Suggest why the data was presented this way.

[1 mark]

....

....

....

Describe and explain the relationship between metabolic rate per gram per hour and body mass.

[3 marks]

....

....

....

....

....

0 7 . 3 Smaller organisms, such as the protist amoeba, have no special tissues, organs or systems for gaseous exchange. Mammals are large, multicellular organisms and have more complex systems.

Explain why mammals need such systems whereas single-celled organisms do not.

[2 marks]

0 8   **Figure 9** shows the pressure changes that occur in the left side of the human heart.

Figure 9

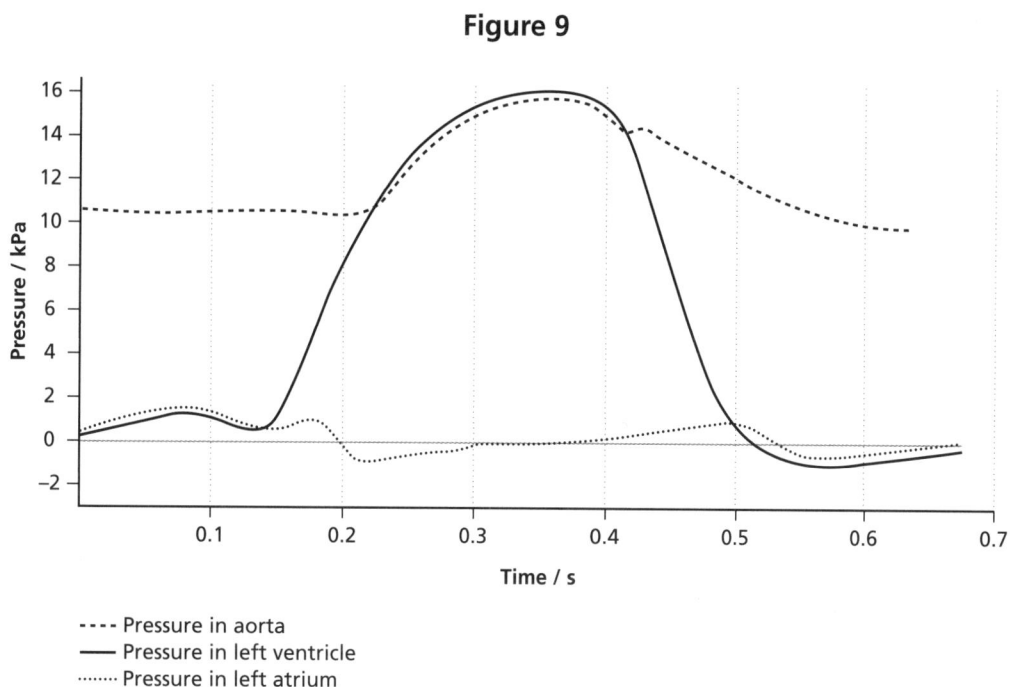

---- Pressure in aorta
—— Pressure in left ventricle
········ Pressure in left atrium

0 8 · 1 From the graph, estimate the duration of a ventricular contraction.

[1 mark]

0 8 · 2 On the graph, mark with an arrow the point at which the semilunar valves shut.

[1 mark]

0 8 · 3 Describe and explain the state of the atrioventricular valves at 0.2 seconds.

[2 marks]

0 8 · 4 Cardiac output can be calculated using the formula:

Cardiac output = stroke volume × heart rate

dm₃ per minute   dm₃ per beat   beats per minute

The resting cardiac output of an athlete is 7.15 dm³ per minute at a heart rate of 65 beats per minute.

Calculate the stroke volume.

[2 marks]

_____ litres

0 8 · 5 After training, the maximum heart rate of an athlete increases by 20 bpm. Explain the advantage of this increase to an athlete.

[3 marks]

_____

_____

_____

_____

_____

0 9   Haemoglobin is a protein found in a wide range of organisms. It can reversibly combine with oxygen to form **oxyhaemoglobin**.

Figure 10 shows the oxygen dissociation curve for human oxyhaemoglobin with normal concentrations of carbon dioxide and with extra carbon dioxide.

Figure 10

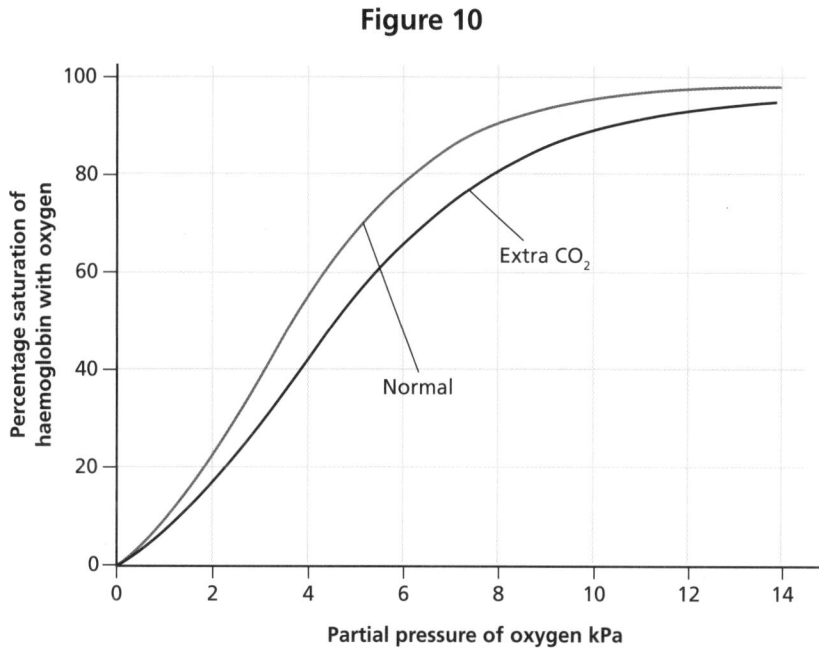

0 9 . 1 Explain how the shape of the normal dissociation curve makes haemoglobin well adapted for oxygen carriage.

[2 marks]

0 9 . 2 Explain the importance of the effect on the oxygen dissociation curve of a high partial pressure of carbon dioxide at a muscle.

[2 marks]

Another protein that acts in a similar way to haemoglobin is **myoglobin**. **Myoglobin** does not travel in the blood but is found in muscle. It has a greater affinity for oxygen than haemoglobin.

0 9 . 3 Draw a curve on **Figure 10** to show the oxygen dissociation curve for myoglobin.

[1 mark]

0 9 . 4 Cetaceans, such as the Sperm whale, dive beneath the ocean surface for long periods of time. They have a very high concentration of myoglobin in their muscles. Explain the advantage of this.

[2 marks]

....

....

....

....

| 1 | 0 |

In the past, farmers were given a subsidy to 'set aside' land previously used for arable farming to encourage the development of natural habitats for a greater range of wildlife. No cultivation was carried out on land set aside in this manner.

Scientists investigated the biodiversity of two equally-sized plots of neighbouring land on a certain farm. One area – **Plot A**, had been set aside for **five** years. The other area – **Plot B**, had been set aside for only **two** years. They used small quadrats of side 0.25 m to sample the two plots and collected data showing the number and range of plant species. Plants were individually counted within the quadrat.

Selected results are shown in **Table 3**.

Table 3

Plant species	Mean number of organisms per quadrat		P value
	Plot A	Plot B	
Common bent	18	2	0.015
Wavy hairgrass	10	25	0.025
Buck's-horn plantain	1	5	0.010
Heath speedwell	0	3	0.001
Rosebay willow herb	11	3	0.067
Harebell	1	3	0.020
Thistle	5	1	0.015

Species diversity is given as:

$$d = \frac{N(N-1)}{\Sigma n(n-1)}$$

where N = total number of organisms of all species

and n = total number of organisms of each species.

| 1 | 0 |·| 1 | Calculate the species diversity for **Plot A** and **Plot B**.

[2 marks]

Plot A species diversity ....

Plot B species diversity ....

1 0 · 2 Suggest reasons for the difference between the species diversity in the two areas.

[2 marks]

....

....

....

....

1 0 · 3 A statistical test was carried out that enabled the scientists to see if the difference in number of organisms in **Plot A** and **Plot B** of each species was significant or not. The **P values** obtained are shown in **Table 3**.

Explain the conclusions that can be drawn from this analysis.

[3 marks]

....

....

....

....

....

....

....

1 1 Describe the structure and function of the molecule ATP.

[5 marks]

[Extra space]

1 2 Compare and contrast the processes of **transpiration** and **translocation** in a plant.

[6 marks]

[Extra space]

1 3 Describe and explain the uses of vaccines in protecting **individuals** and **populations**.

[4 marks]

[Extra space]

END OF QUESTIONS

THIS PAGE HAS DELIBERATELY BEEN LEFT BLANK

Collins

A-level
Biology
Practice paper for AQA

Paper 2   Time allowed: 2 hours

Materials

For this paper you must have:
- a ruler with millimetre measurements
- a scientific calculator.

Instructions
- Use black ink or black ball-point pen.
- Fill in the box at the bottom of this page.
- Answer **all** questions.
- Show all your working.

Information
- The marks for the questions are shown in brackets.
- The maximum mark for this paper is 91.

Name: ....

0 1   Scientists recreated Calvin's classic experiment to investigate the biochemical pathway of the light-independent reaction of photosynthesis.

Figure 1 shows the apparatus used.

Figure 1

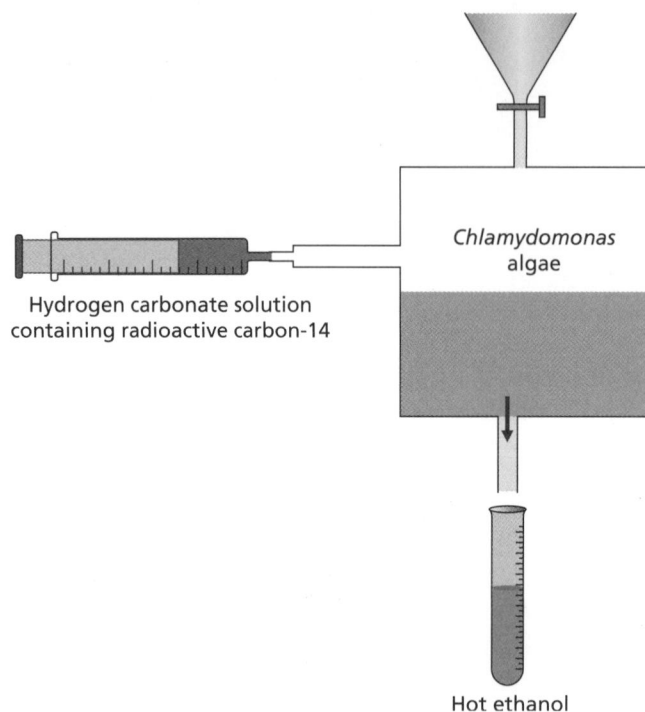

Hydrogen carbonate solution
containing radioactive carbon-14

Chlamydomonas algae

Hot ethanol

The apparatus was placed in a darkened room and the *Chlamydomonas* alga given radioactive carbon-14. The contents of the tank were then mixed and a bright light switched on. Every five seconds, *Chlamydomonas* cells were poured into hot ethanol. The cells were then homogenised (mashed up to form a liquid) and the process of two-way chromatography carried out. This involved performing chromatography with one solvent running, then turning the chromatogram through 90° and running it again with a different solvent. The separated substances had no colour, but the scientists were able to detect the compounds by exposing the chromatogram to a photographic plate. Radioactive compounds were revealed where the plate 'fogged' over.

0 1 . 1 Why was hydrogen carbonate solution introduced into the apparatus?

[1 mark]

0 1 . 2 Why were the algae poured into hot ethanol?

[1 mark]

....

....

0 1 . 3 Explain why the cells were homogenised.

[1 mark]

....

....

Figure 2 shows the results of the two-way chromatography analysis.

Figure 2

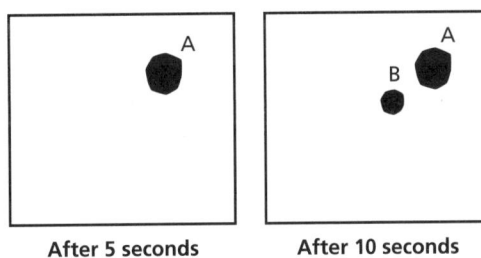

After 5 seconds   After 10 seconds

0 1 . 4 Using what you know of the light-independent stage of photosynthesis, suggest names for compounds **A** and **B**.

[2 marks]

A ....

B ....

Another experiment was carried out to determine the quantities of two compounds produced during the light-independent reaction. The experiment consisted of a light and then a dark period. **Figure 3** shows a graph of the results obtained.

Figure 3

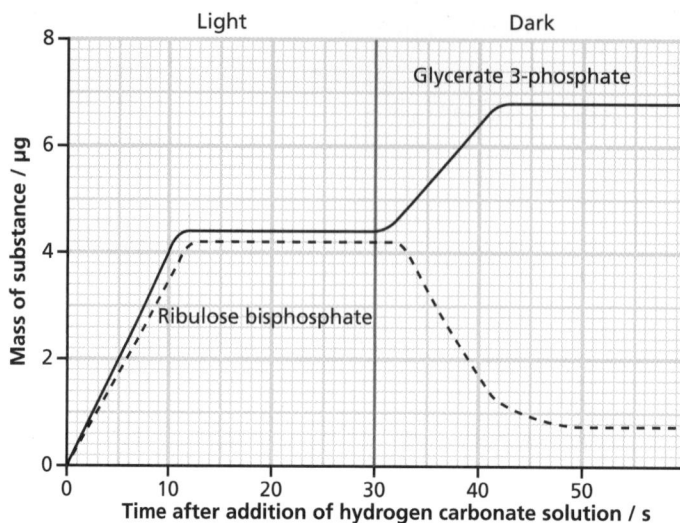

0 1 . 5 Calculate the percentage increase in glycerate 3-phosphate between 30 and 43 seconds to one decimal place.

[2 marks]

.... %

0 1 . 6 Explain why the levels of glycerate 3-phosphate increase after the light is switched off.

[2 marks]

....

....

....

0 1 . 7 Explain why levels of ribulose bisphosphate decrease.

[2 marks]

....

....

....

0 1 . 8 Name the enzyme that catalyses the conversion of ribulose bisphosphate to glycerate 3 phosphate.

[1 mark]

0 2 **Figure 5** shows the main stages of the nitrogen cycle.

Figure 5

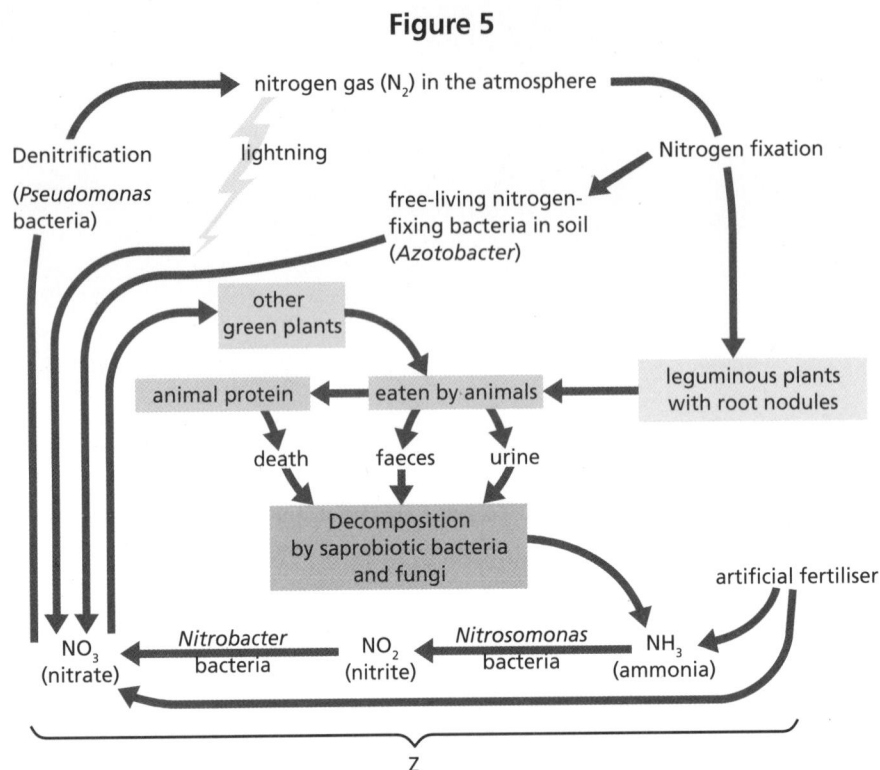

0 2 . 1 Process Z includes the reactions that convert ammonia to nitrate. What is the name of this process?

[1 mark]

....

0 2 . 2 In nitrogen fixation, bacteria such as *Rhizobium* make use of atmospheric nitrogen to make it available for leguminous plants to use. Describe this process, stating the ion initially formed and the compounds in the plant requiring the nitrogen.

[2 marks]

....

....

....

0 2 . 3 Waterlogged soil is very oxygen-poor.

Explain, using your knowledge of denitrification, why such soil also has low concentrations of **nitrate**.

[2 marks]

0 2 . 4 Carnivorous plants such as Venus's flytrap and sundews are able to grow in swamps and bogs despite the soil being nitrogen-poor.

Suggest how they are able to survive.

[2 marks]

0 2 . 5 As well as nitrate, phosphate is an important mineral ion required by plants. Farmers apply fertilisers that contain high concentrations of nitrate and phosphate to fields. When it rains, some of this fertiliser can run off into nearby rivers.

Why do plants require phosphate for healthy growth?

[1 mark]

Scientists sampled the water in seven rivers and streams within the same county. They measured the concentration of phosphates in the water of each river and the mass of algal cells present. Their data are shown in **Table 1**.

Table 1

River	Phosphate concentration gl⁻¹ (x)	Algal concentration µg dm$_{-3}$ (y)	xy	x²	y²
A	343	462	158 466	117 649	213 444
B	198	227	44 946	39 204	51 529
C	317	407	129 019	100 489	165 649
D	243	374	90 882	59 049	139 876
E	161	222	35 742	25 921	49 284
F	193	375	72 375	37 249	140 625
G	195	230	44 850	38 025	52 900
Σ	1650	2297	576 280	417 586	813 307

The table also shows the totals needed in order to calculate a correlation coefficient.

$\boxed{0\,2}\cdot\boxed{6}$ Calculate the **correlation coefficient** (r) for this data using the formula below:

$$r = \frac{n(\Sigma yx) - (\Sigma x)(\Sigma y)}{\sqrt{[n\Sigma x^2 - (\Sigma x)^2][n\Sigma y^2 - (\Sigma y)^2]}}$$

where **n** is the sample size (i.e. 7).

Use the space below to show your working.

[3 marks]

r = ...

0 2 . 7 The scientists concluded that increased phosphate concentrations were causing an increased quantity of algae in the waterways. Do you agree with them? Explain your response.

[1 mark]

Decision (agree/disagree) ....

Explanation ....

....

....

0 2 . 8 The addition of phosphate to waterways can cause **eutrophication**. Describe how this type of pollution harms aquatic habitats.

[3 marks]

....

....

....

....

....

....

The choice chamber in **Figure 6** can be used to investigate the behaviour of the crustacean woodlouse *Armadillidium* and their preference for dark or light conditions.

Figure 6

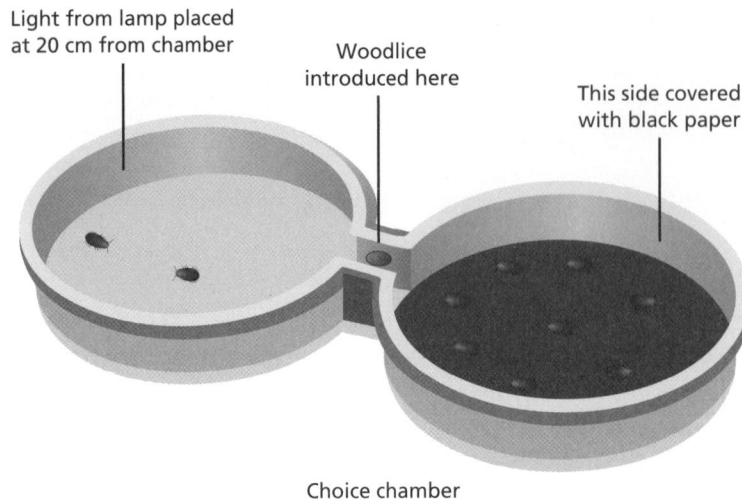

Light from lamp placed at 20 cm from chamber

Woodlice introduced here

This side covered with black paper

Choice chamber

- One side is darkened by covering and securing it with tape, the other is left exposed to a source of artificial light.

- Ten woodlice are introduced to the centre of the choice chamber and the lid is replaced.

- After 5 minutes the number of woodlice are counted in the exposed half. This number is subtracted from the total to find the number of woodlice in the dark half.

- The whole process is repeated 10 times.

Some data from the experiment are shown in **Table 2**.

Table 2

Trial	Woodlice count after 5 minutes	
	In the light	In the dark
1	2	8
2	4	6
3	3	7
4	4	6
5	3	7
6	4	6
7	5	5
8	6	4
9	3	7
10	5	5
Totals		

0 3 · 1 State a suitable null hypothesis for this experiment.

[1 mark]

....

....

0 3 · 2 The students conducting the experiment ensured that they used woodlice of the same species and size. State **two** other variables that should be controlled.

[2 marks]

Variable 1 ....

Variable 2 ....

0 3 · 3 A chi-squared analysis can be carried out to find out whether there are any significant differences shown in the woodlice distribution. Use the experimental data in **Table 2** along with **Table 3** and the formula below to calculate a value of chi-squared.

Table 3

	Light	Dark
Observed results (O)		
Expected results (E)		
$(O–E)^2$		
$(O–E)^2/E$		

$$\chi^2 = \Sigma \frac{(O - E)^2}{E}$$

O = the frequencies observed
E = the frequencies expected
Σ = the 'sum of'

[3 marks]

$\chi^2 =$ ....

0 3 · 4 The students concluded that woodlice show a definite preference for dark conditions. **Table 4** shows critical values. Use **Table 4** to comment on the students' findings.

[2 marks]

Table 4

Degrees of freedom	Critical value for 95% confidence
1	3.84
2	5.99
3	7.82
4	9.49
5	11.07
6	12.59
7	14.07
8	15.51
9	16.92
10	18.31

Degrees of freedom = number of categories (N) − 1

....

....

....

0 3 · 5 Woodlice also exhibit responses to tactile (touch) stimuli. They are said to show a negative taxis to touch. Explain what this term means.

[2 marks]

....

....

0 4   Dihybrid inheritance refers to genetic crosses for genes at different loci (positions) on chromosomes. In pea plants, the colour and texture of seeds are determined by two genes at different loci. Letters are given to the various genes as follows:

R = round seeds (dominant)   r = wrinkled seeds (recessive)

Y = yellow seeds (dominant)   y = green seeds (recessive)

A cross was carried out by a gardener where a heterozygous pea plant for both traits was pollinated by another heterozygous plant. The gardener took the seeds from this cross and counted the number of each type. Her results are shown in **Table 5**.

Table 5

Seed description	Number
Round and yellow	217
Round and green	80
Wrinkled and yellow	66
Wrinkled and green	22

0 4 · 1 Use the genetic cross framework below to show how these progeny could have resulted.

Gametes produced by each parent plant:

....   X   ....

[1 mark]

[2 marks]

0 4 · 2 What ratio would you expect for the seed genotypes in this cross?

[1 mark]

0 4 . 3 Give **two** reasons why the actual numbers obtained from dihybrid crosses may vary from the expected numbers.

[2 marks]

Figure 7 shows a family tree where some individuals have the condition haemophilia.

Figure 7

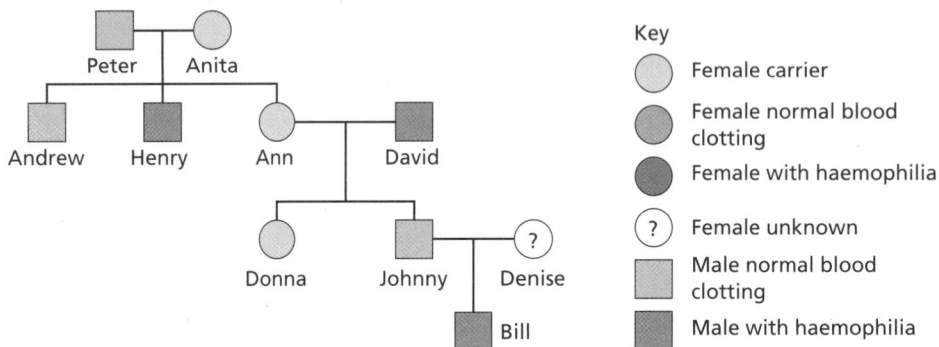

0 4 . 4 Show the **two** possible genotypes for Denise and explain how this can be deduced.

Give evidence from the genetic diagram to support your answer.

Let X^H = normal blood clotting

Let X^h = haemophiliac trait

[1 mark]

Denise's possible genotypes

Explanation

[2 marks]

0 5   Type I diabetes is a condition, usually starting in childhood, whereby the body is unable to regulate its blood sugar levels.

0 5 . 1 Describe the causes and symptoms of Type I diabetes.

[4 marks]

....

....

....

....

....

The treatment of Type I diabetes includes the injection of insulin. A new development is a machine that acts as a continuous glucose monitor and an insulin pump. The device calculates the exact dosage of insulin required, based on the levels recorded. The insulin is then delivered automatically into the skin tissue. See **Figure 8**.

Figure 8

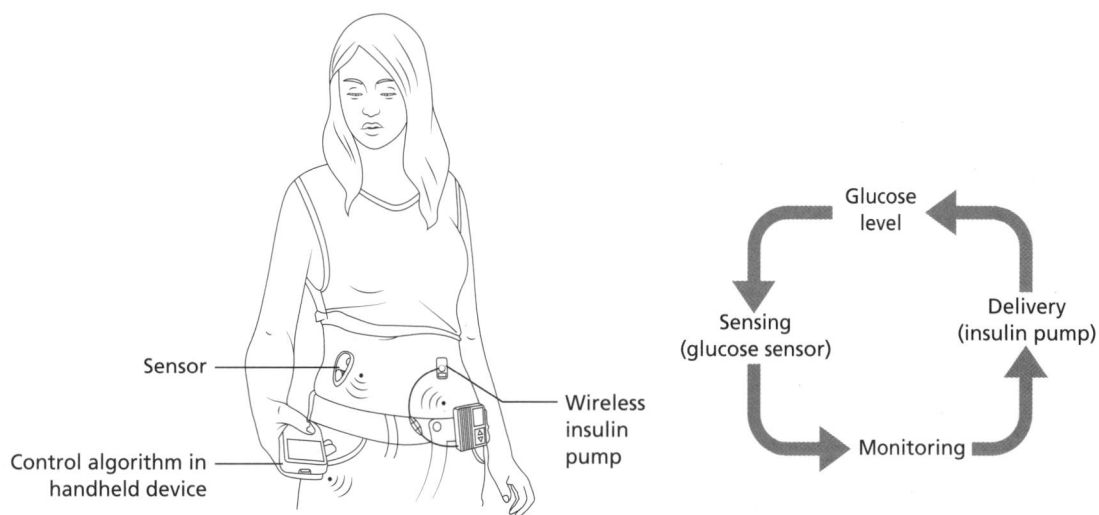

0 5 · 2 What are the advantages and disadvantages of using this method of insulin delivery instead of injections?

[4 marks]

A scientist took a blood sample from a patient suspected of having diabetes two hours after a meal and separated the plasma from the cells using centrifugation. They then made up some standard solutions of glucose. The plasma and glucose samples were then tested with Benedict's reagent. Tubes of each sample were placed in a colorimeter and the light absorbance recorded. The results for the standardised glucose samples are shown in **Figure 9**.

Figure 9

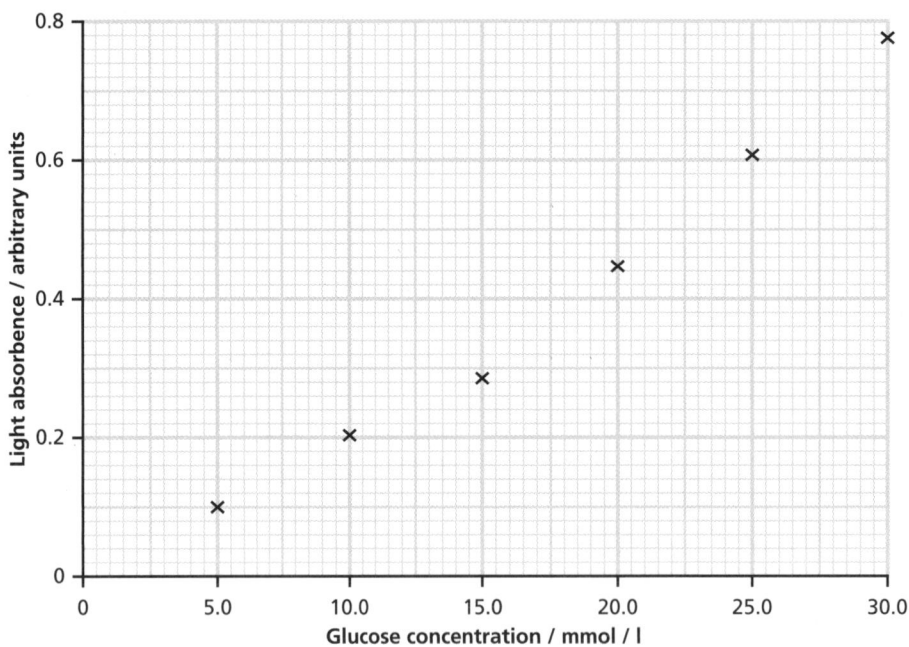

0 5 · 3 The sample from the patient had an absorbency of 0.5 arbitrary units.

Using the graph above to help you, estimate the blood glucose concentration of the patient.

[1 mark]

0 5 · 4 Most adult blood glucose concentrations are below 7.0 mmol/l. What conclusions can you draw about the health of the patient? Give a reason for your answer.

[2 marks]

_____

_____

_____

0 6   A scientist humanely trapped some mice in a small woodland. Using the mark-release-recapture method, she estimated there were 140 mice altogether in the area. Some mice had a dark brown coat and others a light brown coat. Genetic studies show that the allele for dark brown (B) is dominant over the allele for light brown (b).

In the population there were 42 light brown mice.

0 6 · 1 Using the Hardy–Weinberg equations, calculate the frequency of the recessive allele in the population.

[2 marks]

0 6 · 2 Using the Hardy–Weinberg equations, calculate how many heterozygotes there would be in a population of 3000 mice.

[3 marks]

0 6 · 3 Name **two** assumptions that must be made before applying the Hardy–Weinberg principle.

[2 marks]

| 0 | 7 | Read the passages below.

(1) Finches from the Galapagos Islands have become one of the most popular representatives of Darwin's theory of natural selection. They embody the process of speciation forced by environmental conditions. Differences between very closely related species depend on the particular island where they originated, and the food types present there. Studies have continued on these species, and their genomes scrutinised at the molecular level in order to try to understand how exactly speciation occurs. They illustrate the process of allopatric speciation forced by environmental conditions.

(2) A recent study was carried out by a scientific team led by M. K. Skinner. The team compared the pattern of copy number variations (CNVs) between the genomes of different finch species. CNVs are repetitions of genetic sequences that can be related to phenotypic variation and speciation processes.

(3) Skinner also chose a particular kind of modification called DNA methylation. Methylation often switches genes off and can be transmitted to offspring, and therefore can be an important factor in evolution. The question Skinner asked was: which kind of modifications in the finches' genomes showed a pattern that could correlate with the phenotypic differences observed in Darwin's finches? When comparing DNA or protein sequences, the study of their similarities and differences at the molecular level reflects millions of years of evolution from their common ancestors.

Based on information from http://mappingignorance.org/2014/12/01/epigenetics-takes-us-back-galapagos/

| 0 | 7 | · | 1 | The term **allopatric speciation** is used in paragraph 1. Explain the difference between allopatric and sympatric speciation.

In each instance use an example other than Darwin's finches to illustrate your answer.

[4 marks]

0 7 · 2 The terms **genetic sequences** and **genome** are used in the text. Both involve studying the molecule DNA. How would the data produced from these two types of study differ?

[2 marks]

0 7 · 3 Suggest how CNVs (paragraph 2) might be used to identify common ancestors and the pattern of speciation.

[2 marks]

0 7 · 4 What property of the methylation process makes it useful for studying evolution?

[2 marks]

The diagram (**Figure 10**) shows a possible evolutionary tree for several species of Galápagos finches.

Figure 10

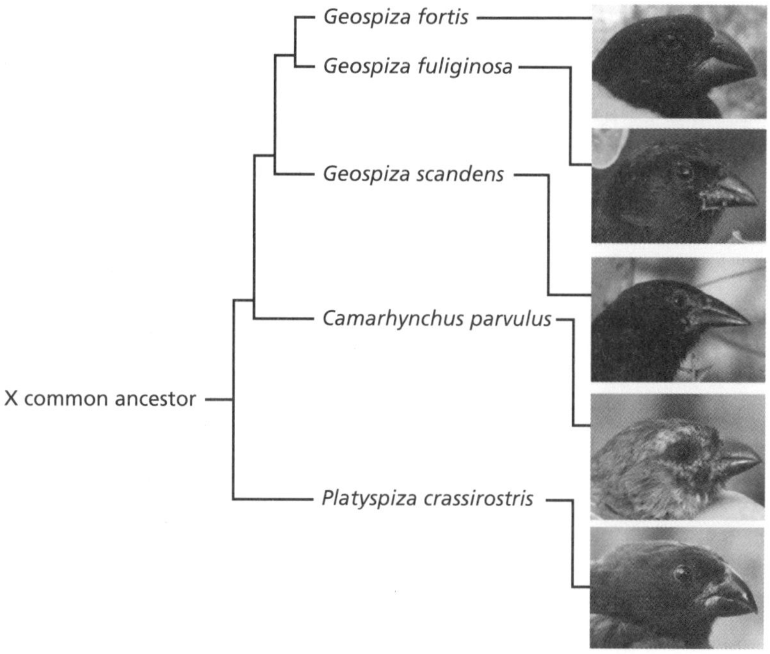

A finch's beak is adapted to the type of food it eats. A larger beak is better for consuming larger nuts, whereas a smaller, thinner beak is more advantageous for dealing with smaller seeds.

0 7 . 5 Species X is termed a common ancestor. Use Darwin's theory of natural selection along with what you know about speciation to explain how the common ancestor could have given rise to the five species shown.

[5 marks]

....

....

....

....

....

....

....

....

0 8　Genes can be inserted into crop plants to produce varieties better suited to meet the demands of food supply. For example, Golden Rice contains a gene that produces beta-carotene. Beta-carotene can be converted to vitamin A in the human body. Lack of vitamin A in the diet can lead to a condition called xerophthalmia. The bacterial vector used to insert the coding gene is *Agrobacterium tumefaciens,* a plant pathogen that causes tumour-like swellings, called galls, to grow.

Figure 11 shows the biotechnological process used to produce a transgenic plant.

Figure 11

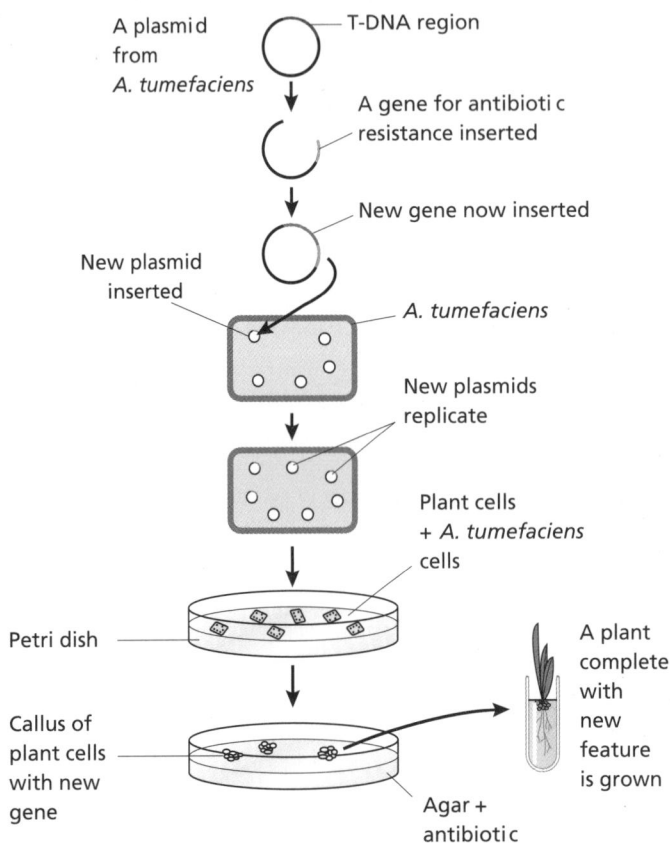

0 8 · 1　In stage 1 of this procedure, a section of the pathogenic bacterial DNA (the T-DNA region) is removed. The discarded code produces proteins that interfere in a plant's production of cytokinins and auxins. Suggest how the removal of this DNA prevents pathogenic action of the bacterium.

[1 mark]

0 8 · 2 What type of enzymes are used to remove genes in this way?

[1 mark]

....

0 8 · 3 Explain how the insertion of a gene coding for antibiotic resistance ensures that only plants containing vitamin A are produced.

[2 marks]

....

....

....

0 8 · 4 The plant cells used in stage 2 have their cell walls removed. Why is this necessary?

[1 mark]

....

....

0 8 · 5 A symptom of xerophthalmia is 'night blindness.' A pigment called rhodopsin is derived from vitamin A. Explain how a lack of vitamin A could lead to night blindness.

[2 marks]

....

....

Figure 12 shows the structure of rod cells taken from a human retina.

Figure 12

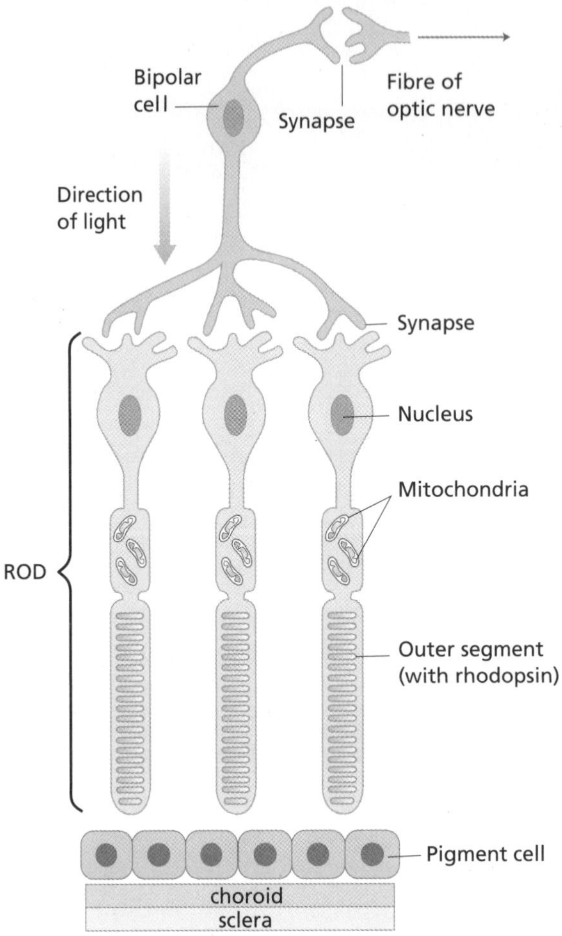

Use the diagram to explain how bright light causes the rod cell to initiate an action potential in an axon from the optic nerve.

[5 marks]

END OF QUESTIONS

Collins

A-level
Biology
Practice paper for AQA

Paper 3   Time allowed: 2 hours

Materials
For this paper you must have:
- a ruler with millimetre measurements
- a scientific calculator.

Instructions
- Use black ink or black ball-point pen.
- Fill in the box at the bottom of this page.
- Answer **all** questions in Section **A**.
- Answer **one** question from Section **B**.
- Show all your working.

Information
- The marks for the questions are shown in brackets.
- The maximum mark for this paper is 78.

Name: _____

Section A

Answer **all** questions in this section.

---

0 1   **Figure 1** shows a simplified biochemical pathway for glycolysis and the link reaction in mammals.

Figure 1

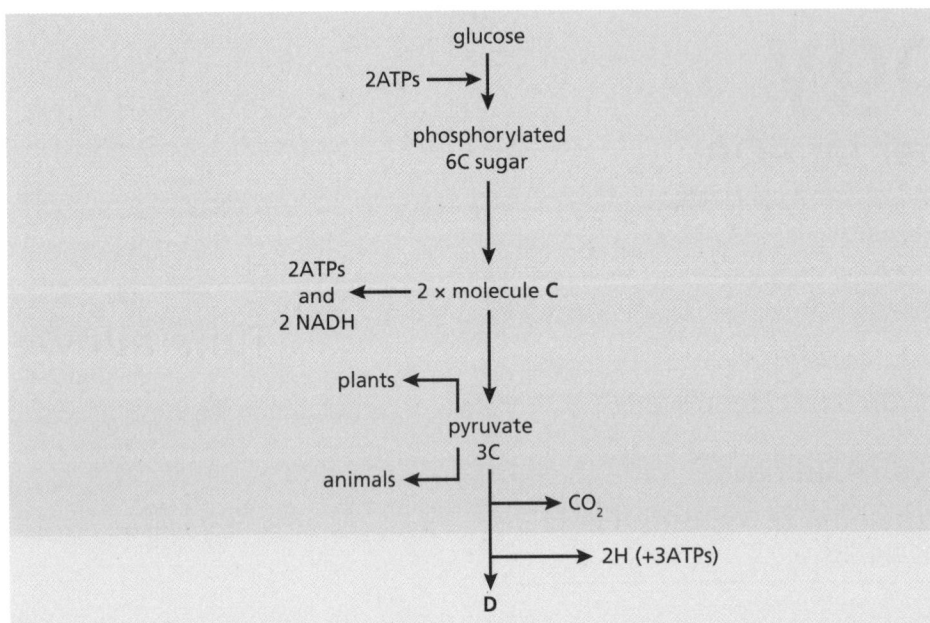

0 1 . 1 Name molecules **C** and **D**.

[1 mark]

....

....

0 1 . 2 In aerobic respiration, reduced coenzyme produced in glycolysis (NADH) is reduced later in the mitochondria and recycled as NAD once more. Explain why glycolysis would stop if the NADH was not used in this way.

[1 mark]

....

....

....

0 1 . 3 During anaerobic respiration, NAD is recycled in a different manner. Describe the fate of NADH in plants and animals undergoing anaerobic respiration.

[1 mark]

Plants

....

....

[1 mark]

Animals

....

....

0 1 . 4 In mammals, prolonged exercise can lead to muscle cramp. Explain why this occurs.

[2 marks]

....

....

....

....

0 2 Insulin is a polypeptide hormone produced in pancreatic tissue. Below is part of the DNA base sequence found in the gene for the polypeptide:

GCA   TAT   AGA   CCA   TCT   GAA   ACA   CTG   TGC   GGC

Figure 2 shows a matrix that can be used to identify the particular amino acid coded for by a triplet of bases in the DNA template.

Figure 2

Second organic base

First organic base	A	G	T	C	Third organic base
A	AAA AAG ⌉ Phe AAT AAC ⌉ Leu	AGA AGG AGT AGC ⌉ Ser	ATA ATG ⌉ Tyr ATT *Stop* ATC *Stop*	ACA ACG ⌉ Cys ACT *Stop* ACC Trp	A G T C
G	GAA GAG GAT GAC ⌉ Leu	GGA GGG GGT GGC ⌉ Pro	GTA GTG ⌉ His GTT GTC ⌉ Gln	GCA GCG GCT GCC ⌉ Arg	A G T C
T	TAA TAG ⌉ Ile TAT ⌉ TAC Met	TGA TGG TGT TGC ⌉ Thr	TTA TTG ⌉ Asn TTT TTC ⌉ Lys	TCA TCG ⌉ Ser TCT TCC ⌉ Arg	A G T C
C	CAA CAG CAT CAC ⌉ Val	CGA CGG CGT CGC ⌉ Ala	CTA CTG ⌉ Asp CTT CTC ⌉ Glu	CCA CCG CCT CCC ⌉ Gly	A G T C

0 2 . 1 Use the information to deduce the amino acid sequence for this section of the polypeptide. The first amino acid has been identified for you.

[1 mark]

Arg   ___   ___   ___   ___   ___   ___   ___   ___   ___

0 2 . 2 DNA is replicated during the cell cycle prior to a mitotic division. Describe the events that occur in DNA replication. State the names of any other enzymes involved.

1 The DNA begins to unwind under the influence of the enzyme DNA helicase.

2 .... **[1 mark]**

....

3 .... **[1 mark]**

....

4 .... **[1 mark]**

....

Cells that undergo mitosis are usually controlled by certain signals. Oestrogen is a growth factor and acts as a regulator for cell division. Sometimes, human breast tissue cells arise that do not possess oestrogen receptors.

Cancer cells do not show **contact inhibition**. This is where cells stop dividing due to signals sent from neighbouring cells.

0 2 . 3 Using this information, explain how a cell without oestrogen receptors can become a cancerous tumour.

[3 marks]

....

....

....

....

....

Genetic sequencing experiments have led to treatments for **melanoma**, a form of skin cancer. A new **kinase inhibitor** drug targets the gene that signals the growth of new blood cells in melanoma tumours.

Figure 3 shows how the blood cells are normally manufactured.

Figure 3

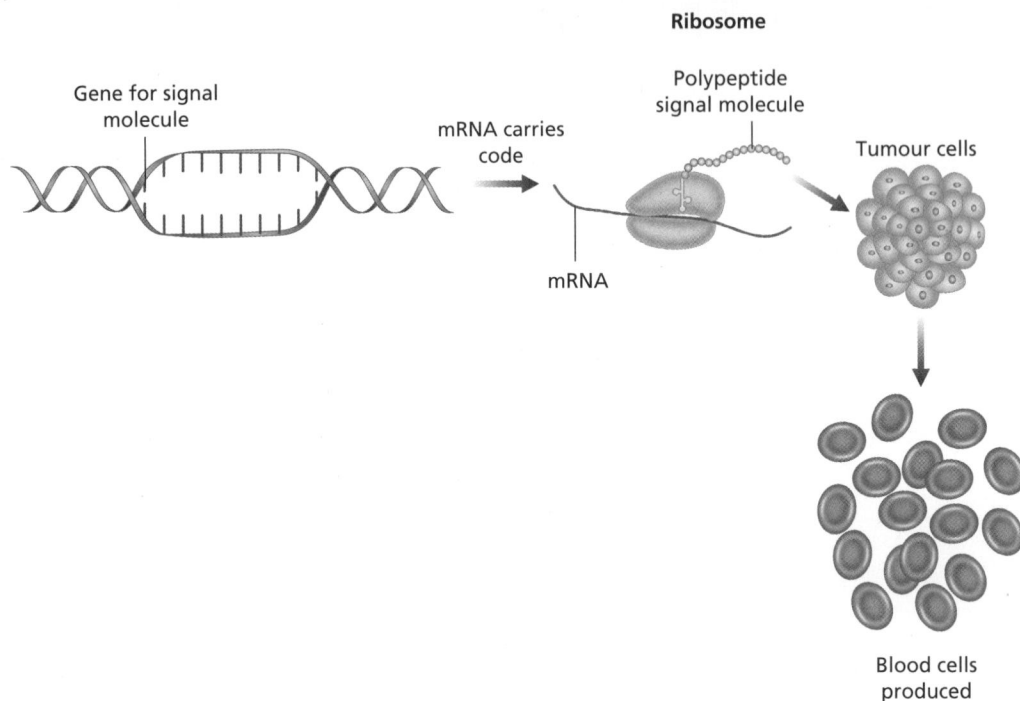

Blood cells produced

0 2 . 4 Using this information, and your knowledge of protein synthesis, explain how the drug can stop the growth of cancerous tumours.

[4 marks]

0 3 **Figure 4** shows how the release of a neurotransmitter called serotonin occurs at a synapse. Drugs called selective serotonin re-uptake inhibitors (or SSRIs) are used in the treatment of depression. Serotonin binds to receptor sites on the post-synaptic membrane and triggers the transmission of impulses. These impulses stimulate areas of the brain associated with the cardiovascular system, muscles, and mood. SSRIs act on serotonin transport molecules on the pre-synaptic membrane.

Figure 4

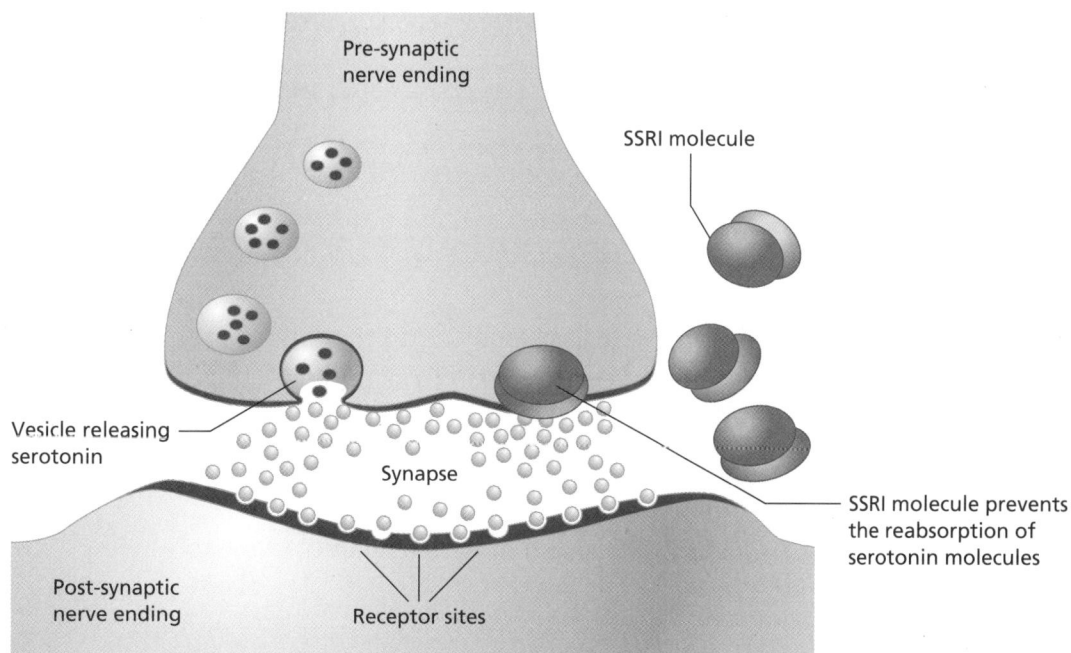

0 3 · 1 Explain how SSRI molecules are able to reduce symptoms of depression.

[4 marks]

Parkinson's disease occurs commonly in the elderly. It results from the degeneration of neurones that produce dopamine. Dopamine is a neurotransmitter that binds to specialised dopamine receptors on the post-synaptic membrane. This binding triggers the generation of action potentials, resulting in lifted mood or even euphoria, depending on the amount of dopamine present. It is also involved in the control of muscle contraction.

0 3 . 2 One type of drug contains a molecule precursor of dopamine. Suggest how this drug would benefit a patient suffering from Parkinson's disease.

[2 marks]

0 3 . 3 Methylenedioxymethamphetamine (MDMA) is a drug that stimulates vesicles in dopamine-releasing neurones and also blocks re-uptake transporter molecules on the pre-synaptic membrane. Explain on a **molecular** level how MDMA produces feelings of well-being.

[2 marks]

0 3 . 4 Explain the importance of summation at a synapse.

[2 marks]

0 4   **Figure 5** shows the structure of an antibody in the class known as **immunoglobulin** G or IgG.

Figure 5

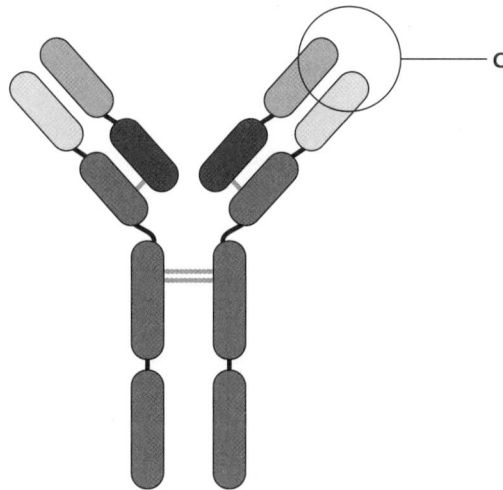

0 4 · 1 Name the site on the antibody labelled **C**.

[1 mark]

....

0 4 · 2 Describe how the site labelled **C** is involved in defending the body against invading pathogens.

[2 marks]

....

....

....

Viperids are a family of snakes that produce venom containing polypeptide toxins. The polypeptide is often an enzyme that coagulates blood – a haemotoxin. When a victim is bitten, it is sometimes difficult to identify the species of snake responsible and therefore administering the correct anti-venom can be problematic.

Scientists have developed a technique for developing polyvalent anti-venoms. These are effective against a range of snake venoms.

Figure 6 shows how snake venom can be identified.

Figure 6

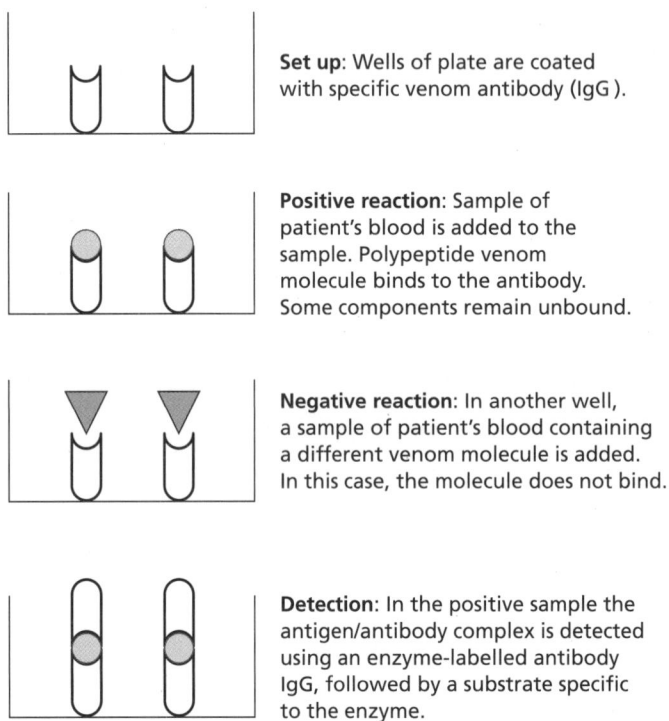

Set up: Wells of plate are coated with specific venom antibody (IgG).

Positive reaction: Sample of patient's blood is added to the sample. Polypeptide venom molecule binds to the antibody. Some components remain unbound.

Negative reaction: In another well, a sample of patient's blood containing a different venom molecule is added. In this case, the molecule does not bind.

Detection: In the positive sample the antigen/antibody complex is detected using an enzyme-labelled antibody IgG, followed by a substrate specific to the enzyme.

0 4 . 3 In the detection stage, the substrate used changes colour in the reaction. How could this be used to determine the amount of venom in the patient's blood?

[2 marks]

0 4 . 4 Scientists have developed a technique for developing polyvalent anti-venoms. These are effective against a range of snake venoms.

Explain how the technique shown in **Figure 6** could be used to develop a **polyvalent anti-venom**.

[2 marks]

0 4 . 5 Using what you know about the tertiary structure of proteins, suggest how venom coagulates the patient's blood.

[2 marks]

0 4 . 6 Describe what happens in a phagocyte after a pathogen is engulfed.

[2 marks]

Plastic debris has been accumulating in marine habitats for decades, to the extent that estimated annual build-up measures in millions of tonnes. Large, persistent congregations of plastic items are quite visible. What isn't as obvious, is the hidden build-up of microscopic fragments and fibres, together with microbeads, a commonly used component in exfoliating products such as toothpastes. See **Figure 7** and **Figure 8**.

Figure 7

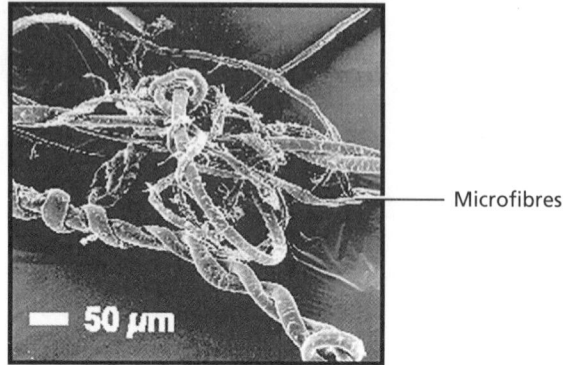

Microfibres

Figure 8

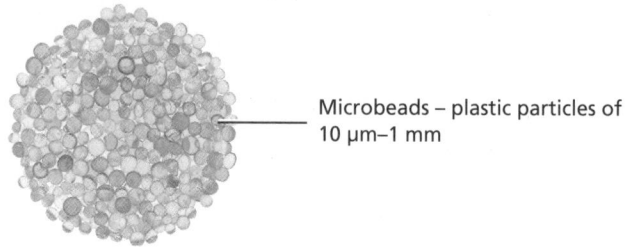

Microbeads – plastic particles of 10 μm–1 mm

To quantify the abundance of microplastics, scientists carried out an investigation.

Sediments were collected from 18 beaches and from their associated estuarine and sub-tidal sediments.

Particles were separated by flotation to remove natural particulate material.

The findings are shown in **Figure 9**.

Figure 9

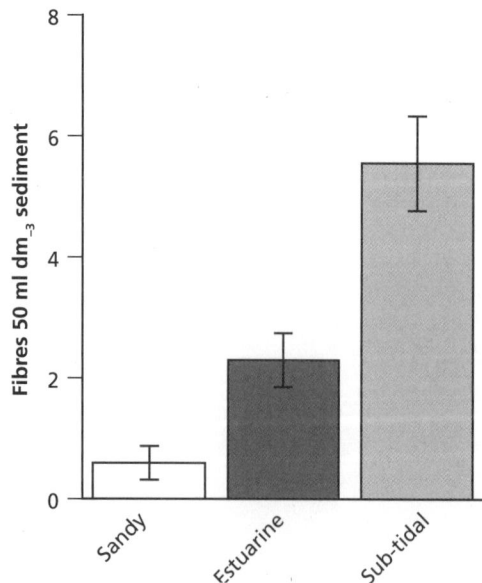

0 5 · 1 What property of the sediment particles is utilised by the separation procedure?

[1 mark]

....

0 5 · 2 What conclusions can be drawn from the data in **Figure 9**?

[2 marks]

....

....

....

To assess long-term trends in abundance, data from plankton samples collected since the 1960s were analysed. The findings are shown in **Figure 10**.

Figure 10

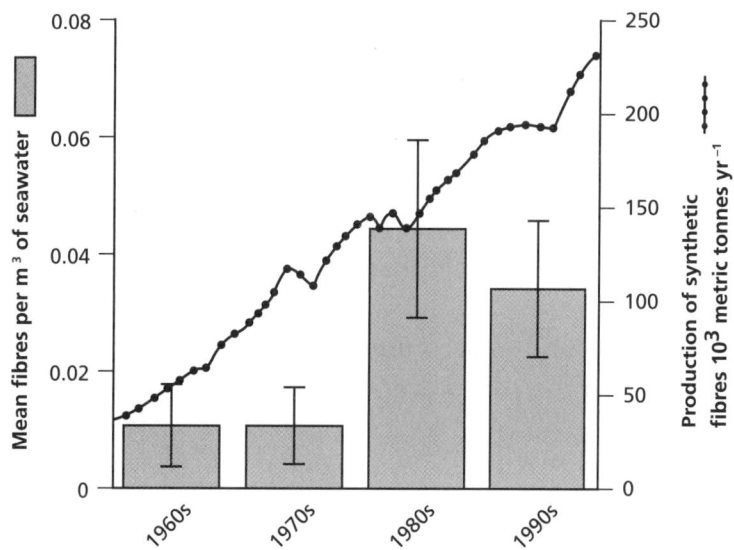

0 5 · 3 What conclusions can be drawn from the data in **Figure 10**?

[3 marks]

....

....

....

....

....

To determine the potential for microscopic plastics to be ingested, the scientists kept amphipods (detritivores), lugworms (deposit feeders), and barnacles (filter feeders) in aquaria with small quantities of microscopic plastics. All three species ingested plastics within a few days.

0 5 · 4 Suggest why the scientists used these organisms for their experiment.

[2 marks]

0 5 · 5 In their paper, the scientists said '. . . it remains to be shown whether toxic substances can pass from plastics to the food chain.' Using information from **Figure 9** and **Figure 10**, comment on their statement.

[2 marks]

Another study of plastics pollution looked at how vertebrates were affected by microbeads. It showed that plastic polymers are 'sponges' for toxins such as DDT and chemicals called PCBs. Microbeads were found to accumulate up to one million times the concentration of these toxins compared to the concentration floating in the water itself.

0 5 · 6 Explain why high concentrations of these toxins might affect vertebrates more than the invertebrates found lower in the food chain.

[2 marks]

Table 1 shows the concentrations of three types of PCB found in different organs of the fish *O. mossambicus*.

Table 1

PCB levels in *O. mossambicus* tissues ($n = 9$, ng·g^{-1}, mean ± standard deviation)				
Analytes	**Muscle (*n*=9)**	**Gills (*n*=9)**	**Gonads (*n*=9)**	**Liver (*n*=9)**
PCB 28	6.47 ± 6.94	7.89 ± 5.48	9.45 ± 10.60	15.26 ± 6.57
PCB 52	3.42 ± 2.33	7.97 ± 4.93	7.97 ± 4.66	17.73 ± 6.34
PCB 101	6.38 ± 4.54	8.34 ± 6.77	8.50 ± 5.49	16.90 ± 3.13

0 5 · 7 A student looked at the data and concluded that PCBs accumulate more in fish livers than in any other organ. Explain why this conclusion may be unreliable.

[3 marks]

....

....

....

....

....

Section B

Answer **one** question.

---

0 6   Write an essay on **one** of the topics below.

EITHER

0 6 · 1 The effects of global warming on the distribution and internal processes of organisms.

[25 marks]

OR

0 6 · 2 The mechanisms by which substances are transported within cells and across the cell surface membrane.

[25 marks]

THIS PAGE HAS DELIBERATELY BEEN LEFT BLANK

Answers

Page 7

1. a) A = nucleotide **[1]**
   It contains a phosphate atom **[1]**
   B = monosaccharide **[1]**
   It only contains carbon, oxygen and hydrogen atoms **[1]**
   C = amino acid **[1]**
   It contains nitrogen but no phosphorus / It contains a sulfur atom **[1]**
   b) $C_6H_{12}N_2O_3S_2$ **[1]**

2. a)
   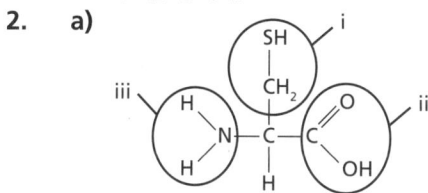
   [3]

   b)
   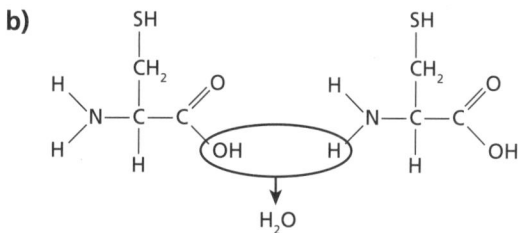
   [2]

3. a)
   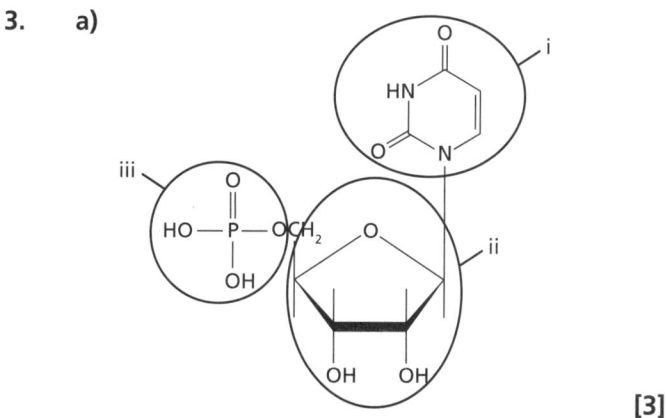
   [3]

   b) Contains a uracil base, which is not found in DNA. **[1]**
   The sugar contains two OH groups rather than one. **[1]**

Page 9

1. A **[1]**
2. a)
   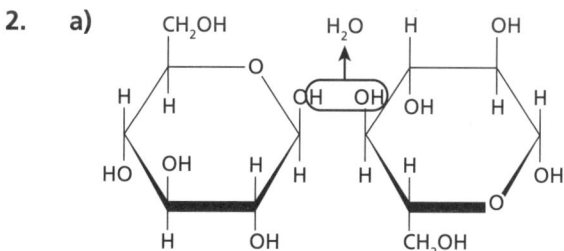
   [2]

   b) β 1-4 bond **[1]**

c) In maltose, the bond is an α bond / maltose has two α-glucose units **[1]**
so one of the glucose units is upside down / the CH_2OH groups are on opposite sides in cellobiose. **[1]**

3. a) $C_{12}H_{22}O_{11}$ **[1]**
   b)
   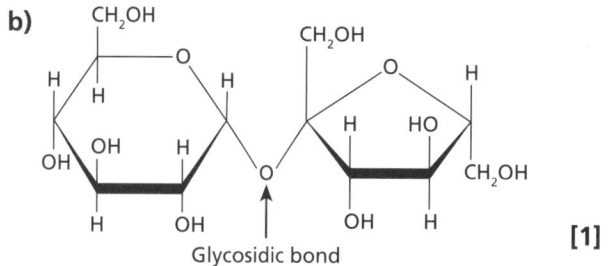
   Glycosidic bond **[1]**

   c) α-glucose + fructose **[1]**

   d) i)
   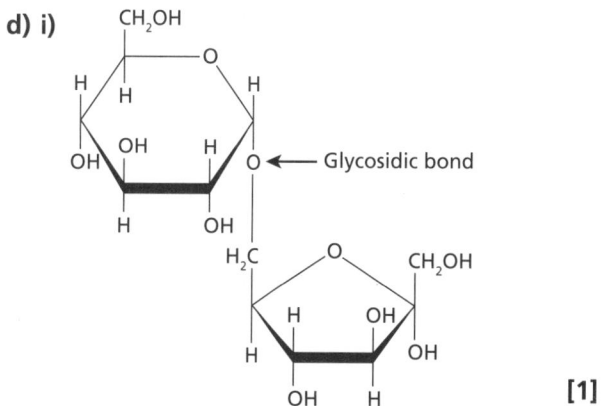
   Glycosidic bond **[1]**

   ii) They have the same molecular formula. **[1]**
   iii) In sucrose, the bond is between OH groups attached to carbon atoms in the ring. **[1]**
   In isomaltulose, one of the OH groups is on a carbon atom that is not in the ring. **[1]**

Page 11

1. One mark for each correct row **[4]**

Feature	Starch	Glycogen	Cellulose
Produced by plants	✓	✗	✓
Contains β-glucose monomers	✗	✗	✓
Contains glycosidic bonds	✓	✓	✓
Forms straight fibres	✗	✗	✓

2. 1 and 4 only **[1]**
3. a) Branched molecule **[1]**
   b) The glycosidic bonds between the monomers are a mixture of α1-4 and 1-6 bonds **[1]**
   The 1-6 bonds produce a branch in the molecule **[1]**

c) Can be broken down quickly [1]
   As there are many sites for enzymes to attack [1]
4. a) β1-4 bonds [1]
   b) Carbohydrates contain C, H and O only [1]
      Chitin also contains N [1]
   c) i) A long, straight molecule [1]
      ii) All the bonds are 1-4, so no branches [1]
         The β bonds mean that each monomer is rotated by 180 degrees so the side groups alternate, forming a straight molecule [1]

Page 13

1. A = a carbohydrate [1]
   Because it has twice as many hydrogen atoms as oxygen [1]
   B = a triglyceride [1]
   Because it has a low proportion of oxygen atoms to hydrogen atoms [1]
   C = a phospholipid [1]
   Because it contains a phosphorus atom [1]
2. 1 and 3 only [1]
3. a) A = Glycerol [1]
      B = Unsaturated fatty acid [1]
      C = Saturated fatty acid [1]
      D = Ester bond [1]
   b) Large amounts of energy are contained in the molecule. [1]
      It is insoluble so will not draw water into cells / will not leak out of cells. [1]
      It is compact so large amounts can be stored in a small space. [1]
4. a) The heads of the phospholipids are on the outside and are attracted to water. [1]
      The hydrocarbon tails are repelled by water and are hidden inside. [1]
   b) Micelles can be used to transport lipids. [1]
      The lipids are trapped on the inside, next to the hydrocarbon tails and away from water. [1]

Page 15

1. a) 3 [1]   b) X [1]   c) 7 [1]
2. a) 1 [1]   b) 2 and 4 [1]   c) 3 [1]   d) 7 [1]
3. a) α-helix [1]
   b) Secondary [1]
   c) Hydrogen bonds [1]
      between hydrogen and oxygen atoms [1]
   d) Hydrogen bonds are weak interactions [1]

Page 17

1. a)

Reagents	Molecule tested for	Observation	Positive or negative result
iodine solution	starch	orange colour	negative
biuret reagent	protein	mauve/lilac colour	positive
Benedict's solution	reducing sugar	red precipitate	positive
ethanol and water	lipids	white emulsion	positive

[8]

   b) It only shows the presence of reducing sugar, not the quantity. [1]
   c) The yogurt is white coloured so it is difficult to see the emulsion. [1]
2. a) Amylose [1]
   b) Orange to blue-black [1]
   c) Iodine needs to become part of the helical structure. [1]
      Glycogen molecules do not form a helix. [1]
   d) The hydrogen bonds will break [1]
      releasing the iodine ions. [1]
3. Add Benedict's solution and heat. [1]
   Filter the liquid to remove the red precipitate. [1]
   Heat the filtrate with hydrochloric acid [1]
   Neutralise with sodium hydroxide solution. [1]
   Repeat the Benedict's test again and there should be a further red precipitate. [1]

Page 19

1. a) Amylase is released outside cells. [1]
   b) The energy of amylose molecules = E [1]
      The energy of maltose molecules = F [1]
      The activation energy of the reaction without amylase = D [1]
      The activation energy of the reaction with amylase = B [1]
   c) B is smaller than D [1]
      Lowers the activation energy for the reaction [1]
   d) F is smaller than E [1]
      The energy of the products is lower than the reactants [1]
2. a) P = substrate [1]
      Q = active site [1]
      R = enzyme-substrate complex [1]
      S = products [1]
   b) In the induced-fit model, the active site changes shape [1]
      to allow the substrate to fit in [1]
3. a) The products of a reaction (ethanal) can fit back into the active site [1]
      Can be converted back into the substrate [1]

b) The two molecules have a similar shape [1]
Can both fit into the active site [1]

Page 21

1. Most enzyme-substrate complexes will be formed at a pH of 7.8 and a temperature of 37°C [1]

2. a) i) Maltose concentration [1]
   ii) More collisions will occur between maltose and maltase molecules. [1]
   Therefore, more enzyme substrate complexes will form. [1]
   b) i) Maltase concentration [1]
   ii) Increasing maltase concentration increases the rate from B to C. [1]
   c) Any one from: pH; temperature; maltase concentration [1]

3. 2 only [1]

4. Both PABA and sulfonamides have similar molecular structure. [1]
Both will enter the active site of the enzyme. [1]
Sulphonamides act as a competitive inhibitor. [1]
Less folic acid is made so bacteria reproduction is reduced. [1]

Page 23

1.

	DNA	RNA
1	Contains deoxyribose sugar	Contains ribose sugar
2	Contains thymine	Contains uracil
3	Two strands	One strand

One mark for each correct row [3]

2. 3 only [1]

3. a) See diagram below [1]
   b) See diagram below [4]
   c) See diagram below [2]

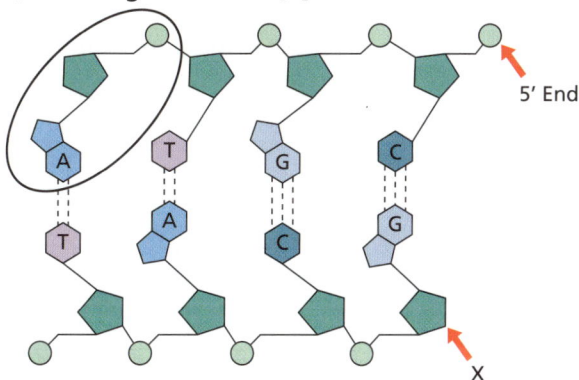

5' End

X

   d) 3' [1]

4. a) Approximately 19.85 for each [2]
   b) Values for A and T are very similar as are G and C [1]
   due to complementary base pairing in the molecule [1]

Human and chicken results are more similar as they are more closely related to each other than to bacteria. [1]

Page 25

1.

	DNA helicase	DNA polymerase
Breaks hydrogen bonds	✓	✗
Breaks phosphodiester bonds	✗	✗
Forms phosphodiester bonds	✗	✓
Unwinds DNA	✓	✗

[8]

2. a) $3.4 \times 10^{-10} \times 6.2 \times 10^9$ [1]
   = 2.11 m [1]
   b) 6 hours = 21 600 seconds [1]
   $(6.2 \times 10^9) \div (50 \times 21\ 600)$ [1]
   = 5.74×10^3 [1]

3. a) i) See diagram below. [1]
   ii) Each new molecule has one new strand and one old strand. [1]
   One strand is light and one strand is heavy. [1]
   Therefore, overall mass is intermediate. [1]

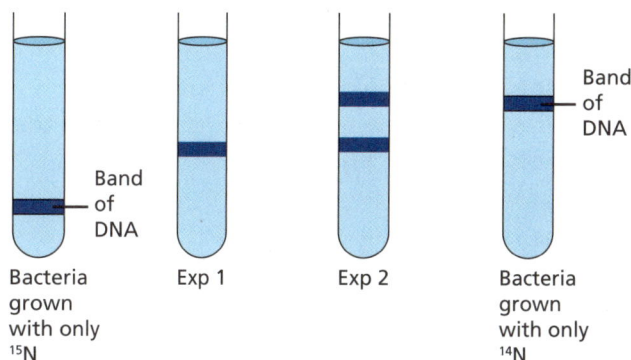

Band of DNA

Bacteria grown with only ^{15}N

Exp 1

Exp 2

Band of DNA

Bacteria grown with only ^{14}N

   b) i) See diagram above. [1]
   ii) Each intermediate molecule unzips and the old light strand combines with a new light strand. The old heavy strand combines with a new light strand. [1]
   Therefore, two types of molecules are made, one completely light and one intermediate. [1]
   So, two bands are formed, one level with the heavy and the other intermediate. [1]

Page 27

1. Photosynthesis and respiration [1]
2. None of the statements [1]

3. a) A = ADP [1]
    B = Water [1]
    C = Adenine [1]
    D = Ribose sugar [1]
    E = Phosphate [1]
    b) See diagram below [2]
    c) See diagram below [2]

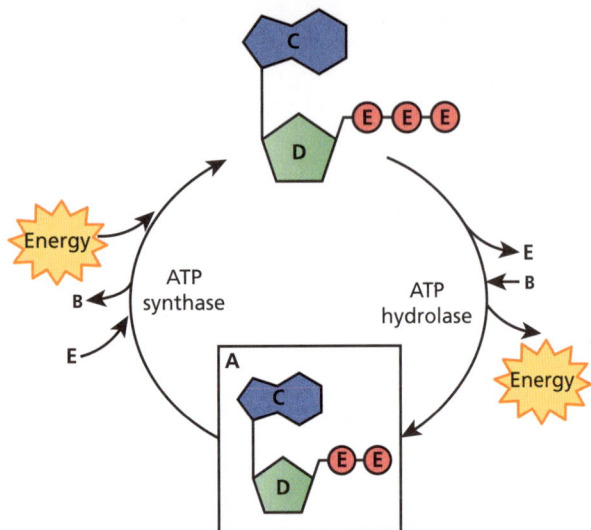

    d) 30.6 kJ mole^{-1} [1]
4. a) $\frac{70000}{50}$ [1]
    1400 : 1 [1]
    b) ATP is not stored in cells. [1]
    It acts as energy currency / It transfers energy. [1]
    It is broken down to release energy. [1]
    It is broken down at about the same rate that it is produced. [1]

Page 29

1. 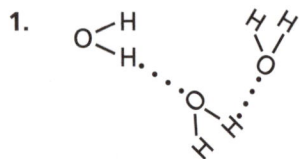 [2]

2. It is required to convert dipeptides to amino acids. [1]

3.

Observation	Property of water
Sea temperature shows less variation than air temperature	Cohesion
Glucose is carried dissolved in blood plasma	High specific heat capacity
Water can be drawn 100 m up the trunk of a tree	High latent heat
Evaporation of water from leaves cools plants	Polar molecule

    All correct [3]; two correct [2]; one correct [1]

4. Covalent bonds involve sharing of electrons. [1]
    Hydrogen bonds involve attraction between partially charged atoms. [1]

Covalent bonds are within the molecule. [1]
Hydrogen bonds are between the molecules. [1]

5. a) pH is higher outside cells [1]
    due to a higher H^+ concentration [1]
    b) Na^+ is higher in concentration outside cells [1]
    Glucose can enter coupled with Na^+ [1]
    c) Positively charged sodium ions will diffuse into the neurone. [1]

Page 31

1. 2 and 4 only [1]
2. 1 and 4 only [1]
3.

Structure	Feature
Lysosome	Transport of lipids
Golgi apparatus	Site of ribosome production
Smooth ER	Contains hydrolytic enzymes
Nucleoli	Modifies and packages proteins

    All correct [3]; two correct [2]; one correct [1]

4. a) A = Rough ER [1]
    B = Golgi apparatus [1]
    D = Ribosome [1]
    b) Radioactive threonine is incorporated into proteins. [1]
    Protein synthesis occurs on the rough ER (A) so radioactivity appears here first. [1]
    Proteins move to the Golgi apparatus (B) so this becomes radioactive next. [1]
    Proteins are modified and packaged. [1]
    Secretory vesicles carry the protein to the cell membrane and so are radioactive next. [1]

Page 33

1.

	Prokaryotic cells	Eukaryotic cells	Viruses
Contain ribosomes	✓	✓	✗
Contain cytoplasm	✓	✓	✗
Have a protein coat	✗	✗	✓
Contain plasmids	✓	✗	✗
Have a cell wall made of murein	✓	✗	✗

    One mark for each correct row [5]

2. B [1]
3. Glycoproteins are used to attach viruses to their host cells [1]
    so that they can enter/inject their genetic material [1]
4. a) Mitochondria and chloroplasts [1]

b) They are formed by an infolding of the cell membrane. **[1]**
The inner membrane comes from the cell that is engulfed and the outer one from the cell that takes up the prokaryote. **[1]**

c) Most ribosomes in eukaryotic cells are in the cytoplasm. **[1]**
The prokaryotic ribosomes stay in the organelles. **[1]**

d) Viruses need a host cell to reproduce. **[1]**
If they came before living cells, they would not have been able to reproduce. **[1]**

Page 35

1. The shortest distance apart that two objects can be seen separately. **[1]**
2. B **[1]**
3. Any two from:
Quicker preparation; less chance of artifacts appearing; can view live specimens; can see colours; cheaper to buy **[2]**
4. a) Transmission electron microscope (TEM) **[1]**
b) Drying / fixing **[1]**
Cut into thin slices **[1]**
Stained **[1]**
c) Diameter = 3 cm = 30 000 μm **[1]**
$6000 = \frac{30\,000}{actual\ size}$ **[1]**
Actual size = 5 μm **[1]**
d) i) Homogenisation/lysis/cell fractionation **[1]**
ii) To inactivate enzymes **[1]**
so that the specimen is not digested/hydrolysed **[1]**
iii) X = nuclei Y = mitochondria Z = ribosomes **[3]**

Page 37

1. The chromosomes gather at the equator of the cell. **[1]**
2. B **[1]**
3. a) A = anaphase **[1]**
B = telophase **[1]**
C = metaphase **[1]**
D = prophase **[1]**
b) The chromosomes become thicker **[1]**
c) Between metaphase (C) and anaphase (A) **[1]**
4. a) 24 hours = 1440 minutes **[1]**
$\frac{3}{533} \times 1440 = 8.11$ **[1]**
= 8 minutes **[1]**
b) i) $\frac{23}{533}$ **[1]**
= 0.043 **[1]**
ii) Cancer cells divide rapidly **[1]**

Page 39

1. C **[1]**
2. a) Fluid mosaic model **[1]**
b)

The area where cholesterol is found	T
A carbohydrate molecule	R
A peripheral protein	Q
A hydrophobic region	T

1 mark for each correct letter **[4]**

3. a) $4 \times 3.14 \times 25^2$ **[1]**
7850 μm² **[1]**
b) 3.14×75^2 **[1]**
17662.5 μm² **[1]**
c) Area of monolayer is about double that of the surface area of the cell **[1]**
That is because the cell membrane is a bilayer **[1]**

Page 41

1. 2 and 4 only **[1]**
2. B **[1]**
3. a) Facilitated diffusion **[1]**
b) It does not require energy/ATP from respiration **[1]**
as it is moving molecules down a concentration gradient **[1]**
4. a) 55 – 10 = 45 **[1]**
$\frac{45}{10} \times 100$ **[1]**
= 450% **[1]**
b) Plant Y cannot respire **[1]**
so no energy/ATP **[1]**
Active transport of nitrate into the roots cannot occur **[1]**
c) Diffusion **[1]**
d) Concentration of nitrate ions becomes the same inside and outside the root **[1]**
so no gradient for diffusion to occur. **[1]**

Page 43

1. P **[1]**
2. a) 1 **[1]**
b) Statement 1 – the immune system cannot produce proteins quickly enough to bind with the antigens **[1]**
Statement 2 – the immune system does not detect the parasite as foreign **[1]**
Statement 3 – the immune system cannot reach the parasite and detect its antigens **[1]**
Statement 4 – phagocytosis cannot occur **[1]**
3. a) So that they can be seen and counted **[1]**
b) 9 – 6 = 3 **[1]**
$\frac{3}{9} \times 100$ **[1]**
= 33.3% **[1]**

c) Membrane has less unsaturated fatty acids so less fluid [1]
Makes it harder to flow around the beads [1]
so less phagocytosis occurs [1]

Page 45

1. Bone marrow [1]
2. 1, 2, 3 and 4 [1]
3. a) 4 [1]
   b) 2 [1]
   c) Can clump together antigens / cause agglutination [1]
   So phagocytes can engulf the antigens [1]
   d) If negative selection did not occur, B cells would produce antibodies against body cells [1]
   Body cells would be destroyed [1]
4. a) P = Antigens [1]
   Q = Antigen-receptor complex / antigen-antibody complex [1]
   R = Antibody [1]
   b) Clonal selection [1]
   c) Mitosis [1]
   d) Needs to produce cell clones/identical cells [1]
   so that all the antibodies made are identical [1]
   e) T helper cell / TH cell [1]

Page 47

1. A fetus gaining antibodies across the placenta [1]
2. Antigens bind to antibodies on the test line. [1]
3. a) Maximums are 300 and 140 [1]
   2.14 : 1 [1]
   b) It lakes longer for the antibody levels to rise after the first dose. [1]
   c) After the second dose, there are memory cells in the blood. [1]
   They produce antibodies quicker and in larger amounts. [1]
4. a) As HIV count increases, the T helper cell count decreases [1]
   As HIV reproduces in T helper cells, killing them [1]
   b) Symptoms do not appear until about 8 years. [1]
   So HIV not diagnosed for a long time. [1]

Page 49

1. 3 only [1]
2. a)   [5]

Organism	Surface area	Volume	SA : Vol
A	314	523	0.6 : 1
B	600	1000	0.6 : 1
C	62	30	2.1 : 1

b) Organism C has the largest SA : Vol [1]
so is more likely to be able to exchange enough gases [1]
Organisms A and B have the smallest SA : Vol so may need extra structures for exchange [1]

3. a) Water is lost through spiracles [1]
   In hot/dry conditions, water needs to be conserved [1]
   b) Open every 20 seconds [1]
   so 3 times a minute [1]
   c) Open for 4 seconds, so 12 seconds a minute [1]
   so 20% [1]
   d) Percentage would increase [1]
   More oxygen is needed [1]
   due to increased respiration in muscles [1]

Page 51

1. 3 and 4 only [1]
2. a) 5 [1]
   b) $4 \div 3500$ [1] = 0.001 mm [1]
   c) Walls are close together [1]
   Epithelial and endothelial cells are thin/ squamous [1]
3. a) X drawn at any point where the line is falling [1]
   b) 16 [1]
   c) One breath = 0.4 dm³ [1]
   So PVR = $16 \times 0.4 = 6.4$ dm³ [1]
   d) PVR would increase [1]
   Increased gas exchange required [1]
   More oxygen needed for respiration / more carbon dioxide to excrete [1]

Page 53

1. None of the above [1]
2. a) glucose = cotransport
   monoglycerides = diffusion [1]
   b) Glucose is a polar molecule / monoglycerides are lipid soluble [1]
   Glucose may be moving against a concentration gradient [1]
   c) Maltose is found attached to the cell membranes [1]
   Allows a steep diffusion gradient to build up [1]
3. a) Lipids digested by lipase [1]
   Fatty acids are produced which change the pH [1]
   b) Both tubes contained bile salts [1]
   Lipids were emulsified [1]
   Larger surface area for lipase to work on [1]
   c) Boiling the lipase in tube 4 will denature the enzyme [1]
   Therefore the lipase will not digest the lipid [1]
   Boiling bile salts does not prevent them working, as they are not enzymes [1]

4. a) Cells contain proteins [1]

   If pepsin is active, it could digest the cells [1]

   b) Endopeptidases cut up the long molecules [1]

   so produce more ends for exopeptidases to work on [1]

   c) Dipeptidases [1]

Page 55

1. 8 [1]

2. 3 only [1]

3. a) 88 to 36 [1]

   52% [1]

   b) 88 to 64 [1]

   24% [1]

   c) Bohr shift [1]

   More oxygen is released in areas where carbon dioxide is produced by respiration [1]

   More aerobic respiration / less anaerobic respiration [1]

4. a) At 4 kPa, it is 50% saturated [1]

   so 0.7 ml [1]

   b) At 4 kPa, it is 73% saturated [1]

   so 1.0 ml [1]

   c) more [1]

   affinity [1]

   placenta [1]

Page 57

1. 1 and 3 [1]

2. A = aorta [1]

   B = coronary artery [1]

   C = right ventricle [1]

   D = vena cava [1]

3. a) Pressure is lower on the right side by about four times [1]

   Muscle in the wall of right ventricle is thinner [1]

   Only needs to pump blood to the lungs rather than all around the body [1]

   b)

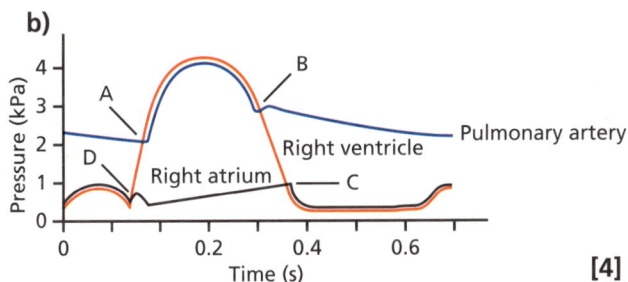

[4]

4. a) $\frac{60}{0.8}$ [1]

   = 75 [1]

   b) $\frac{0.1}{0.8} \times 100$ [1]

   = 12.5% [1]

   c) About 0.4 s [1]

   The ventricles have just stopped contracting [1]

   So pressure in the arteries would be higher than the pressure in the ventricles [1]

Page 59

1. a) P = arteries [1]

   Q = capillaries [1]

   R = venules [1]

   b) 18.5–13.5 kPa [1]

   c) Pressure in Q is lower [1]

   Pressure in Q does not fluctuate [1]

   d) There is elastic tissue in the artery walls [1]

   Produces elastic recoil when the heart relaxes [1]

2. Any three from:

   Narrow lumen reduces speed of blood flow

   More time for exchange

   Thin walls / walls one cell thick

   Less distance for exchange with the tissues [3]

3. a) Difference in hydrostatic pressure = 3.2 [1]

   Difference in water potential = 2.0 [1]

   Therefore, net pressure of 1.2 forcing tissue fluid out of blood [1]

   b) Any two from: glucose; amino acids; oxygen; minerals; vitamins [2]

   c) X = lymph [1]

   Drains into thoracic duct [1]

   Empties into a vein near the heart [1]

Page 61

1.

	Xylem vessels	Sieve tube cells
Living cells	✗	✓
Contains lignin	✓	✗
Transports substances both up and down the stem	✗	✓
Has end walls	✗	✓
Contains some cytoplasm	✗	✓

One mark for each correct row [5]

2. a) When the xylem and phloem are separated, the radioactivity is all in the xylem. [1]

   Activity in the phloem in this area matches background radiation. [1]

   b) Above and below the paper, the radioactivity is almost equal in both xylem and phloem. [1]

Radioactive potassium must be equally distributed in both areas although it is transported in the xylem. **[1]**

   c) Ions can move sideways out of the xylem through pores. **[1]**

3. a) Phloem sieve tubes **[1]**
   b) source = leaves **[1]**
     sink = growing points / storage organs / fruits **[1]**
   c) Water is drawn into the source by osmosis. **[1]**
     This increases the hydrostatic pressure in the source. **[1]**
     Contents of source are forced along the tube to the sink. **[1]**

Page 63

1. 3 only **[1]**

2.

	Eukaryotic nuclear DNA	Mitochondrial DNA	Bacterial DNA
Associated with histone proteins	✓	✗	✗
Codes for all the proteins made in the cell	✗	✗	✓
Contains introns	✓	✗	✗
Enclosed by a double membrane	✓	✓	✗

   One mark for each correct row **[4]**

3. a) $4^3 = 64$ but $4^2 = 16$ **[1]**
     so a double code cannot code for enough amino acids **[1]**
   b) There are 64 possible codes but only 20 amino acids. **[1]**
     Therefore, many of the amino acids are codes for by more than one triplet. **[1]**

4. a) 51 amino acids to code for **[1]**
     so needs $51 \times 3 = 153$ base pairs **[1]**
   b) Locus **[1]**
   c) Exons **[1]**
   d) It codes for insulin that is closer in structure to human insulin. **[1]**
     Therefore, the insulin is more likely to work like human insulin. **[1]**
   e) The genetic code is universal. **[1]**
     Therefore, bacteria will produce insulin with the same structure as human insulin. **[1]**

Page 65

1. 1 only **[1]**

2.

	Transcription	Translation
Involves base pairing	✓	✓
Involves enzymes	✓	✓
Involves DNA	✓	✗
Breaks peptide bonds	✗	✗

   One mark for each correct row **[4]**

3. 2, 4, 1, 5, 3
   One mark for each in the correct order (with two correct to gain the first mark) **[4]**

4. a) 2 **[1]**
   b) UAC **[1]**
   c) Met Arg Phe His **[1]**
   d) They do not code for an amino acid **[1]**
     so translation stops at this point. **[1]**

Page 67

1. 3, 2, 4, 1
   One mark for each in the correct order (with two correct to gain the first mark) **[3]**

2.

	Mitosis	Meiosis
Homologous chromosomes pair up	✗	✓
Spindle fibres shorten	✓	✓
Produces two cells	✓	✗
Reduces the chromosome number	✗	✓

   One mark for each correct row **[4]**

3. a) 6 **[1]**
   b) Metaphase I – showing chromosomes of all sizes divided into two chromatids; **[1]**
     homologous chromosomes lined up in pairs in the centre **[1]**
     Metaphase II – one chromosome from each pair still made of two chromatids **[1]**
     lined up separately at the centre of the cell **[1]**
     Telophase II – chromatids separated from each other; **[1]**
     one from each chromosome at the poles of the cell **[1]**
   c) Cytokinesis will occur. **[1]**
     Cytoplasm will be pitched in at the centre. **[1]**
   d) In metaphase of mitosis, the chromosomes will not be paired up. **[1]**
     They will be lined up separately at the equator. **[1]**

4. a) A = DNA replication **[1]**
     B and C = cytokinesis **[1]**
     D = fertilisation **[1]**
   b) On the horizontal line between B and C **[1]**

Page 69

1. 64 [1]
2. 2 and 3 only [1]
3. a) A = Centromere [1]
   B = Chromatid [1]
   b) Drawings showing one chromosome/ chromatid in each cell with the alleles G and H, G and h, g and H, g and h [4]
4. a) AUGUGGCUCGCU [2]
   b) Met Trp Leu Ala [1]
   c) i) Substitution [1]
      ii) The amino acid Cys would be included rather than Try [1]
      iii) The codon ACT would be translated as UGA; [1]
          this is a stop codon so transcription would stop [1]

Page 71

1. Behavioural [1]
2. Discontinuous variation which is controlled by genes [1]
3. 1 and 3 only [1]
4. C E D A B
   One mark for each pair of letters in the correct order [4]
5. a) Any one from: they were well adapted; lack of predators; plenty of food [1]
   b) A mutation occurs in a rabbit; [1]
      this makes a rabbit resistant to the virus. [1]
      The rabbit does not die, so survives to reproduce [1]
      and passes on the allele for resistance. [1]
      The allele for resistance increases in the population. [1]

Page 73

1. Only one narrow range of phenotype is selected for. [1]
2. a) Stabilising selection [1]
      If there are more spines, cacti are better protected from predators, but if there are too many spines, they are more likely to be parasitised. [1]
      Therefore, an intermediate density is selected for. [1]
   b) Directional selection [1]
      In drought conditions, birds with large beaks are selected for as they can eat the larger seeds. [1]
      They will reproduce and so the next generation will have larger beaks. [1]

c) Stabilising selection [1]
   Short plants will not get enough light and so are more likely to die. [1]
   Very tall plants would need to use large amounts of raw materials for support. Therefore, medium height plants are selected for. [1]
3. a) Total number of eggs = 4 × 36 = 144 [1]
      Percentage hatched = $\frac{108}{144}$ × 100 = 75% [1]
   b) Stabilising selection [1]
   c) Any two from:
      Female geese may not be able to keep them warm.
      They may be targeted more by predators
      They may be too small / not contain enough food reserves. [2]
   d) More eggs might hatch but the young may be less likely to survive. [1]

Page 75

1. a) To make sure that they mate with a member of the same species. [1]
      As they look similar, courtship allows them to identify their own species. [1]
      If they mated with another species then hybrids would be infertile. [1]
   b) Many different species in a small area [1]
      Courtship prevents aggression due to defending territories [1]
   c) Difficult for birds to see each other due to dense forest [1]
2. a) Left column completed from top: Phylum, Genus [2]
      Right column completed from top: *Balaenoptera*, *acutorostrata* [2]
   b) Scientific name is universal / to avoid confusion [1]
      May have different common names in different places [1]
   c) A hybrid [1]
      which would be infertile [1]
   d) i) 12 675 [1]
      ii) Families in common = 10 863 [1]
          $\frac{10\,863}{12\,675}$ × 100 = 86 [1]
      iii) Cow [1]
      iv) Phylogenetics [1]

Page 77

1. None of the above [1]
2. a) Younger fields tend to have a larger area. [1]
      Farmers have removed hedges to make it easier to use large machinery. [1]

b) A = 1.7 **[1]**
   B = 2.0 **[1]**
   C = 1.2 **[1]**
c) Removing any hedges reduces the number of plants. **[1]**
   Old hedges have a higher species richness of plant species. **[1]**
   Removal means fewer habitats / less food for animals. **[1]**

3. **a)** Table completed from left: 6, 7, 4, 6 **[1]**
   b) $N(N-1) = 506$ **[1]**
   Sum of $n(n-1) = 114$ **[1]**
   So $d = 4.44$ **[1]**
   c) Pond B is more stable because it has a higher species diversity **[1]**
   and can therefore adapt to changes more easily. **[1]**

Page 79

1. 3 only **[1]**
   a) To avoid sampling bias **[1]**
   b) Shell diameter = 57.6 mm **[1]**
   Shell height = 21.6 mm **[1]**
   c) There is a greater variation in the height of the shells on the exposed shore. **[1]**
   d) Shell height is shorter and diameter is wider on the exposed shore. **[1]**
   Higher shells are more likely to be knocked off rocks by the waves. **[1]**
   Wider shells can have a wider sucker to hold on to the rocks. **[1]**

3. **a) i)** Gibbon and orangutan **[1]**
   They only have two/the least differences between the amino acids. **[1]**
   ii) Gibbon and gorilla **[1]**
   They have the most/six differences between amino acids. **[1]**
   b) The genetic code is degenerate. **[1]**
   The base sequence may be different but the amino acid codes for the same. **[1]**

Page 81

1. ATP, $NADPH_2$ and O_2 **[1]**
2. None of the above **[1]**
3. NADP **[1]**
4. **a)** D **[1]**
   b) C **[1]**
   c) C **[1]**
   d) D **[1]**
5. When the light is switched on, the pH of the stroma increases. **[1]**
   Light is available to eject electrons from PSI and PSII. **[1]**

More electrons are passed along the electron transfer chain. **[1]**
More protons are pumped out of the stroma into the thylakoids. **[1]**

Page 83

1. In the stroma **[1]**
2. None of the above **[1]**
3. **a) i)** RuBP concentration increases as it is the acceptor of carbon dioxide. **[1]**
   Without carbon dioxide, it cannot be converted to GP. **[1]**
   ii) GP concentration falls as less carbon dioxide is fixed. **[1]**
   Therefore, less RuBP is converted to GP. **[1]**
   b) TP concentration would fall **[1]**
   as there would be less GP to be converted to TP. **[1]**

4. **a)** Rate of carbon dioxide fixation is higher. **[1]**
   Carbon dioxide fixation peaks at a lower temperature. **[1]**
   Carbon dioxide fixation occurs over a wider range of temperatures. **[1]**
   b) In hot climates, oxygen fixation is higher **[1]**
   so it is an advantage to use the different enzyme. **[1]**
   However, in temperate climates there is not as much sunlight **[1]**
   so there is less energy available for the alternative reactions of photosynthesis. **[1]**

Page 85

1. 2 ATP, 2 $NADH_2$ and 2 pyruvate **[1]**
2. 1 and 3 only **[1]**
3. **a)** A = Pyruvate **[1]**
   B = Ethanal **[1]**
   C = Carbon dioxide **[1]**
   b) Ethanol dehydrogenase **[1]**
   c) Lactate is made rather than ethanol. **[1]**
   No carbon dioxide is given off. **[1]**
4. **a)** In the cytoplasm **[1]**
   b) Triose phosphate **[1]**
   c) Lysis: 4 **[1]**
   Phosphorylation: 1 and 3 **[1]**
   Production of ATP: 5 **[1]**
   Production of $NADH_2$: 5 **[1]**

Page 87

1. Pyruvate is oxidised and decarboxylated. **[1]**
2. 3 only **[1]**
3. **a)** Table completed as follows:
   ATP molecules produced directly by the Krebs cycle: 2 **[1]**

Total range of ATP molecules produced by respiration: 36–38 [1]

b) It must be against a concentration gradient / membrane may be impermeable to $NADH_2$. [1]
Active transport is needed. [1]

c) Fewer protons pumped across the inner membrane. [1]
Therefore, smaller proton gradient generated / fewer protons to pass through ATPsynthase. [1]

4. a) Protons were pumped out, across the inner membrane. [1]
They could not accumulate in the intermembrane space / they diffused away. [1]
Smaller proton gradient produced [1]
so fewer protons to pass through ATPsynthase. [1]

b) If permeable then the protons would leak back in when pumped out. [1]
No proton gradient would be set up, so no ATP production [1]

Page 89

1. a) Heat transfer tube is coiled [1]
Lid on top of the apparatus [1]

b) Oxygen is supplied [1]

c) Temperature change = 15°C [1]
energy needed = 15 × 200 × 4.2 = 12 600 [1]
So per g = 630 J [1]

2. a) 65 000 × 10% = 6500 [1]
6500 × 20% × 20% = 260 kJ m^{-2} [1]

b) Energy is lost at each trophic level [1]
Not enough left to sustain another level [1]

c) Less material is indigestible in animals. [1]
Some parts are not eaten in plants. [1]

3. a) J or kJ per unit of area per unit of time [1]

b) i) 87 402 [1]
ii) Primary productivity [1]

c) Respiration [1]

d) 89 [1]

e) It may not be of a suitable wavelength. [1]

Page 91

1. Nitrification requires oxygen. [1]

2. a) Acts as a control to compare with the results of the other treatments [1]

b) i) $\frac{112}{104} \times 100$ [1]
= 107.7% [1]
ii) $\frac{110}{104} \times 100$ [1]
= 105.8% [1]

c) Both pot 2 and pot 3 showed a greater increase in mass of leaves and tubers than pot 1. [1]

Increased leaf growth in pot 2 and 3 led to greater photosynthesis. [1]
Therefore, more carbohydrate production for tuber growth. [1]
Very little difference between the tuber growth in pot 2 and 3. [1]

d) Use of natural fertiliser and mycorrhiza on fields will cause less possible eutrophication than the use of artificial fertilisers. [1]
Difference in the potato yield is not high enough with artificial fertilisers to take the risk. [1]

3. a) Leaching [1]

b) i) Denitrifying bacteria [1]
ii) It converts nitrates to nitrogen gas [1]
so prevents nitrates entering the river. [1]

Page 93

1. Sensory neurone – Relay neurone – Motor neurone [1]

a) More turns on the water side [1]
Movement lines are longer on the water side [1]

b) Kinesis [1]

c) Air above the water is more humid/has more moisture than above the calcium chloride [1]
Woodlouse prefers the air to be humid [1]
and spends more time on the humid side [1]

3.

	Response of shoot	Explanation of response
Shoot 1	No bending/ grows straight up	Light is detected by the tip – tip is covered so IAA evenly distributed
Shoot 2	Grows to the right/light	More IAA is sent from the tip to the shaded side
Shoot 3	Grows to the left	IAA can only reach the side nearest the light
Shoot 4	Grows to the right	IAA sent to the shaded side

One mark for each correctly completed box [8]

4. a) Wide range of concentrations so logarithmic scale allows them to fit. [1]

b) Increases the length of the roots and decreases the length of the roots. [1]
Effect levels out on roots but starts to decrease in shoots. [1]

c) Below that concentration, IAA has no effect on shoots [1]
Below that concentration, IAA stimulates growth in roots [1]
In gravitropism of roots, IAA inhibits growth and stimulates shoots [1]

1.

Feature	Rods	Cones
Produces a generator potential when stimulated by light	✓	✓
Found mainly in the fovea	✗	✓
Contains pigments sensitive to light	✓	✓
Provides high visual acuity	✗	✓
Are sensitive to most wavelengths of light	✓	✗

One mark for each correct row [5]

2. a) X = Lamella [1]
    Y = Axon / Nerve fibre [1]
    Z = Sensory neurone [1]
    b) Generator potential [1]
    c) Deforms the lamella [1]
    Puts pressure on the axon/nerve fibre [1]
    Opens Na⁺ channels causing Na⁺ to enter the axon [1]
    d) High pressure produces a larger generator potential. [1]
    Higher pressure causes the generator potential to last for longer. [1]

3. a) A = rod   B = cone [1]
    b) B (cone) gives higher visual acuity [1]
    as only connected to a single sensory neurone. [1]
    The brain can therefore differentiate between light hitting two different cones. [1]
    c) A (rods) are more sensitive in dim light. [1]
    Several rods are connected to one sensory neurone. [1]
    Light hitting different rods may be enough to stimulate the sensory neurone. [1]

1. a) Bottom box = Cardiac centre / Medulla oblongata [1]
    Box above = SA node [1]
    b) P = Chemoreceptor [1]
    Q = Stretch/Pressure/Mechano receptor [1]
    c) Sympathetic nerve [1]
    d) Pass down the Purkinje fibres [1]
    in the bundle of His. [1]
    Spread up over the ventricular muscle from the base [1]

2. a) AV node [1]
    b) Delay is 0.02 seconds [1]
    Speed = $\frac{0.05}{0.02}$ = 2.5 ms⁻¹ [1]
    c) Pathway from B to C is along Purkinje fibres. [1]

These fibres are large and so transmit rapidly. [1]
    d) This makes the ventricles contract from the base upwards [1]
    so the blood inside is forced upwards into the arteries. [1]

1. Maintenance of a resting potential = X
    Depolarisation = Z
    Repolarisation = Y
    One mark for one correct. [2]

2. a) Resting potential = –70 mV [1]
    Threshold potential = –50 mV [1]
    b) Stimulus is not large enough. [1]
    Potential has not reached the threshold. [1]
    c) Na⁺ channels have opened. [1]
    Na⁺ has entered the axon. [1]
    d) Axon is refractory. [1]
    Na⁺ channels are not able to open again yet. [1]

3. a) Myelinated – as diameter increases so does conduction velocity. [1]
    Graph is a straight line [1]
    Unmyelinated – as diameter increases so does conduction velocity. [1]
    However, not a straight line / rate of increase of velocity decreases. [1]
    b) Action potential jumps from node to node. [1]
    The nodes are the only place that ions can leave or enter the neurone. [1]
    Saltatory conduction occurs. [1]

1. C, B, F, A, E, D
    One mark for each pair of letters in the correct order. [5]

2. a) Vesicles with neurotransmitter are only found in X. [1]
    Receptor sites are only found in Y. [1]
    b) Cocaine blocks the re-uptake sites. [1]
    So, levels of dopamine remain high in synaptic cleft. [1]
    Next neurone remains stimulated. [1]

3. a) Acts as a neurotransmitter [1]
    b) Is a similar shape to acetylcholine [1]
    Competes for the active site of the enzyme [1]
    c) Prevents acetylcholine from being broken down [1]
    So next neurone is continually stimulated [1]
    Insect goes into spasms / cannot fly or move properly [1]

4. a) Temporal summation [1]

b) Cl⁻ entering the postsynaptic membrane will counter the effect of Na⁺ entering. [1]
So, there is less change in membrane potential / less depolarisation. [1]
Action potential less likely to be generated in postsynaptic neurone. [1]

Page 103

1. F, A, C, D, B, E
One mark for each pair of letters in the correct order. [5]

2. 2 and 3 [1]

a) P = Myelin sheath Q = Vesicle
R = Muscle / Postsynaptic membrane
S = Motor end plate [4]

b)

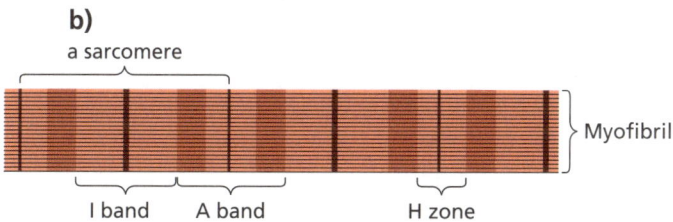

a sarcomere
I band A band H zone
Myofibril

One mark for each. [4]

c)

A band	Stays the same
I band	Gets smaller
H zone	Gets smaller
Sarcomere	Gets smaller

One mark for each correct row. [4]

Page 105

1. a) Set point / Norm [1]

b)

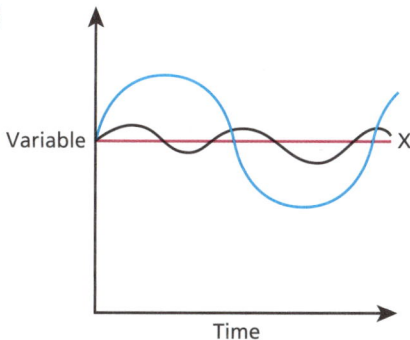

Line still showing fluctuations but smaller than before [1]

2. The level of thyroxine is detected by receptors and is compared to a set point. [1]
If too low then more TSH is released, which stimulates more thyroxine release. [1]
This increases thyroxine levels back to the set point. [1]
Negative feedback [1]

3. a) Receptors = stretch receptors in the cervix [1]
Coordinator = brain [1]
Effector = muscle in the uterus [1]

b) Receptors in the cervix detect the baby pushing against the cervix. [1]
This results in more release of oxytocin, [1]
which causes greater contractions of the uterus, so more stretching. [1]
Positive feedback [1]

Page 107

1. beta; receptors; channel proteins; glycogen; glycogenosis
One mark each [5]

2. a) So their blood glucose at the start is their fasting level [1]

b) The level is 7.0 mmol dm⁻³ [1] which is above the range of 3.9–5.5 mmol dm⁻³ [1]

c) Insulin is still being produced [1]
but blood glucose levels are not being returned to normal. [1]

3. a) Increase = 50 to 640 = 590 [1]
Percentage increase = $\frac{590}{50}$ × 100 = 1180% [1]

b) Insulin [1]

c) Stimulates the production of glycogen [1]
from glucose [1]

d) Glucagon binds with a receptor. [1]
G protein is activated. [1]
Triggers adenylate cyclase to make cAMP. [1]
cAMP activates protein kinase. [1]
By a cascade of reactions activates glycogen phosphorylase. [1]

Page 109

1. In the hypothalamus [1]

2. a) It is zero because it is not filtered out of the blood [1]
because it is too large. [1]

b) The fluid becomes more concentrated because water passes out by osmosis. [1]
This is because of the high concentration in the tissues of the kidney. [1]

c) Water is reabsorbed, [1]
therefore increasing the concentration of the urea. [1]

3. a) Many mitochondria [1]
to produce ATP for selective reabsorption. [1]
Many microvilli/basal folds [1]
to increase surface area for reabsorption. [1]

b) Reabsorption of glucose decreases the water potential of the blood. [1]
So water follows by osmosis. [1]

4. a) A = Glomerulus [1]
B = Bowman's capsule [1]
C = Proximal convoluted tubule [1]
D = Loop of Henle [1]
E = Collecting duct [1]

b) Very long loop of Henle [1]

so very high concentration is built up in the tissues of the kidney. [1]

More water is drawn out of the tubule by osmosis. [1]

Smaller volume of more concentrated urine is produced. [1]

1. 1 red-flowered : 1 white-flowered : 2 pink-flowered [1]

2. **a)**

		Red parent	
		R	r
Red parent	R	Rr	Rr
	r	Rr	rr

One mark for gametes and one mark for genotypes. [2]

b) Breed them with white-flowered plants. [1]

If the offspring are all red then the parent red plant should be homozygous. [1]

If any of the offspring are white then the parent must be heterozygous. [1]

c) i) red = 112.5 [1]

white = 37.5 [1]

ii) $\chi^2 = \frac{(105 - 112.5)^2}{112.5} + \frac{(45 - 37.5)^2}{112.5}$ [1]

$= 0.5 + 1.5$ [1]

$= 2.0$ [1]

iii) One degree of freedom [1]

Critical value is 3.84 [1]

χ^2 is lower than the critical value so accept the null hypothesis / there is no significant difference between the observed and expected ratio [1]

3. **a)** Recessive [1]

Persons 1 and 2 (or 6 and 7) do not have albinism but have children who do have it. [1]

b) i) 4 out of 14 [1]

= 28.6% [1]

ii) 7 out of 14 [1]

= 50% [1]

c) Person 10 is heterozygous / Aa; person 11 is homozygous recessive/aa [1]

Probability of having albinism = $\frac{1}{2}$ [1]

So probability of a boy with albinism = $\frac{1}{4}$ [1]

1. **a)** Black with rough coat = BB/RR, Bb/RR, Bb/Rr, BB/Rr [1]

White with rough coat = bb/RR, bb/Rr [1]

b) i) Black with rough coat = Bb/Rr [1]

White with smooth coat = bb/rr [1]

ii) Parental genotypes = Bb/Rr   bb/rr

Gametes = B/R, B/r, b/R, b/r × b/r [1]

White smooth		Black rough			
		B/R	B/r	b/R	b/r
	b/r	Bb/Rr	Bb/rr	bb/Rr	bb/rr

[1]

Ratio = 1 black rough : 1 black smooth : 1 white rough : 1 white smooth [1]

2. **a)** (1 = X^hX^H), 2 = X^HY, 3 = X^hY, 5 = X^hY, 7 = X^hX^H, 8 = X^HY

One mark for each [5]

b) 50% or $\frac{1}{2}$ or 1 : 1 [1]

c) $X^hX^H \times X^HY$

X^h or $X^H \times X^H$ or Y [1]

X^hX^H X^hY X^HX^H X^HY [1]

= 25% or $\frac{1}{4}$ or 1 : 3 or 1 in 4 [1]

d) Males have only one allele for blood clotting [1]

as they have only one X chromosome. [1]

Can either have the condition or not carrying the allele. [1]

1. **a)** The genes are linked on the same chromosome. [1]

Independent segregation cannot occur. [1]

The gametes T/g and t/G are not produced. [1]

b) i) Crossing over has occurred [1]

to produce the gametes T/g and t/G. [1]

ii) Recombinants [1]

2. **a)** Epistasis [1]

b) Parents = Cc/Ww   cc/ww

Gametes = C/W, C/w, c/W, c/w × c/w [1]

		C/W	C/w	c/W	c/w
	c/w	Cc/Ww	Cc/ww	cc/Ww	cc/ww

[1]

white   coloured   white   white [1]

Expected ratio = 3 white : 1 coloured [1]

3. **a)** Genotype = I^BI^O [1]

Phenotype = blood group B [1]

b) Group A or group B [1]

c) Parents = I^OI^O I^AI^B [1]

Gametes I^O I^A or I^B [1]

Offspring I^OI^A or I^OI^B [1]

Probability = $\frac{1}{2} \times \frac{1}{2} = \frac{1}{4}$ or 25% or 1 in 4 [1]

1. 3 only [1]

2. **a)** Both Rr [1]

b) i) $q = 0.3$ so $q^2 = 0.09$ [1]

$3000 \times 0.09 = 270$ yellow [1]

So red = 2730 [1]

ii) Allele frequency of R will drop and r increase [1] as red-shelled snails more likely to be eaten so pass on less R alleles. [1]

3. a) $q = 0.152$ [1]
    So, $p = 0.848$ [1]
    Therefore, $2pq = 2 \times 0.848 \times 0.152 = 0.2578 = 25.8\%$ [1]

    b) i) In the UK, allele frequencies are likely to be different [1]
       as there is no malaria. [1]

    ii) The allele frequency of the thalassaemia allele is increasing [1]
        as carriers do not get malaria / are less likely to die. [1]

Page 119

1. 1 and 3 only [1]

2. a) Ability to survive [1]
       Ability to reproduce [1]

    b) i) A = Disruptive selection [1]
       The tadpoles with intermediate feeding features are selected against. [1]
       B = No selection [1]

    ii) In crowded ponds there is more competition. [1]
        Specialising in either carnivorous or herbivorous feeding reduces the competition. [1]

    iii) There would need to be an isolation mechanism to prevent them interbreeding. [1]
         Mutations would need to build up in the two separate populations. [1]
         This would mean that they were unable to produce fertile offspring if they did meet and reproduce. [1]

Page 121

1. D [1]

2. a) They do not need to feed on blood to reproduce. [1]
       No birds present in the tunnels to feed from. [1]
       Do not need to hibernate as the temperature does not drop so low in the tunnels. [1]

    b) i) Try to mate the mosquitoes with mosquitoes from above ground. [1]
       If they cannot produce fertile offspring then they are a new species. [1]

    ii) Allopatric [1]

3. a) On the ground, the longer legs allow them to outrun predators. [1]
       Up in the trees, the shorter legs make them more agile. [1]

    b) Disruptive [1]

    c) Sympatric [1]

Page 123

1. a) 3 and 4 only [1]

    b) C [1]

2. a) 1×10^5 [1]

    b) Grass production follows a similar pattern to deer numbers. [1]
       However, the deer numbers peak after the grass does. [1]
       If there is more grass then more deer can survive and numbers increase. [1]

    c) Deer numbers would drop. [1]
       Interspecific competition for food. [1]

3. a) $120 = \dfrac{40 \times 30}{\text{number in second sample}}$ [1]
       Number in second sample = 10 [1]

    b) This could make them more obvious to predators [1]
       which would make the estimate too large. [1]

Page 125

1. The climax community that is formed can vary in different climates. [1]

2. a) Advantages:
       Trees regrow much quicker. [1]
       Can be harvested again much sooner. [1]
       Roots are not removed so less soil erosion. [1]
       Disadvantages:
       Thinner trunks are produced. [1]

    b) More light reaches the floor of the woodland. [1]
       Greater ability for smaller plants to photosynthesise and grow. [1]
       More habitats / More varied food supply for animals. [1]

3. a) Pioneer species = Lyme grass, Dominant species = Deciduous trees [2]

    b) Incorporated organic matter [1]

    c) Light intensity falls between 100 and 200 years ago [1]
       corresponds to increase in trees that would shade the ground. [1]

    d) Greater variety of plants present before the climax. [1]
       Combined productivity of the system is greater. [1]

Page 127

1. a) i) Substitution of T for A. [1]
       Addition of A before T. [1]
       Deletion of A if next base is T. [1]

    ii) mRNA is now UUU [1]
        so amino acid is Phe. [1]

    b) i) Lys becomes Gln as the second amino acid. [1]

ii) Could change the tertiary structure of the enzyme. [1]
Substrate might not fit into the active site. [1]
iii) A deletion mutation has occurred. [1]
Each triplet downstream could change. [1]
Would change many more amino acids. [1]

2. a) Hydrophilic amino acids tend to be attracted to the periphery of the molecule [1]
so are likely to change the tertiary structure of the enzyme. [1]
Particularly important at that point because amino acid 40 is part of the active site. [1]

b) i) Substitution mutation [1]
C changed to G in the codon UCU [1]
ii) Disulfide bond is a stronger bond than hydrogen bonds [1]
so is less likely to be damaged by higher temperatures [1]
and will maintain the shape of the active site. [1]

Page 129

1. They are produced from body cells. [1]

2.

Feature	Type of stem cell		
	Totipotent	Pluripotent	Multipotent
Can be isolated from adult tissues	✗	✗	✓
Can form placental cells	✓	✗	✗
Have some of their genes switched off	✗	✓	✓
Can form all the cells in the human body	✓	✓	✗

One mark for each correct row [4]

3. a) Multipotent [1]
b) 0.0035 – 0.0003 [1]
= 0.0032 [1]
$\frac{0.0032}{30}$ = 0.000107 [1]
c) Bones take longer to heal in older people. [1]
This is because there are fewer stem cells in the bone marrow in older people. [1]
So new cells are produced at a slower rate. [1]
d) Any four from:
Using embryonic stem cells raises more ethical issues.
This is because embryos are destroyed.
Mesenchymal stem cells are easier/quicker to obtain.
Can come from the patient so less risk of rejection.

However, embryonic stem cells can produce a wider range of cell types.
So can treat a wider range of conditions. [4]

Page 131

1. a) F, E, A, D, C, B
One mark for each pair in the correct order [5]
b) Insulin is a protein but oestrogen is a steroid. [1]
Steroids are lipid soluble. [1]
c) i) Oestrogen works by stimulating the production of new proteins. [1]
ii) Responses to adrenaline need to be rapid [1]
to prepare the body for fight or flight/action. [1]
Producing a new protein would take too long. [1]

2.

Feature	Process		
	Use of transcription factors	Epigenetics	RNA interference
Destroys RNA polymerase	✗	✗	✗
Involves enzymes	✓	✓	✓
mRNA is destroyed	✗	✗	✓
Involves methylation	✗	✓	✗

One mark for each correct row [4]

3. a) Transcription [1]
b) siRNA is binding with mRNA [1]
by complementary base pairing [1]
c) RISC [1]
d) The gene is not expressed [1]
because the mRNA cannot be translated [1]
so a protein is not produced. [1]
e) siRNA that is complementary to viral RNA is produced. [1]
This is delivered into cells. [1]
Viral RNA is destroyed so new viruses cannot be produced. [1]

Page 133

1. 1, 2, and 3 [1]
2. a) Oncogene [1]
b) Produces DNA methyltransferase. [1]
This adds methyl groups to the promoters of the tumour suppressor gene. [1]
This gene cannot be transcribed. [1]
c) Histones do not allow the DNA to unwind [1]
so tumour suppressor genes cannot be transcribed. [1]

3. a) Androgens cannot be converted to oestrogen. **[1]**
   Less oestrogen to bind with oestrogen receptors. **[1]**
   Less transcription of oncogenes. **[1]**
   b) Lower levels of oestrogen in the body. **[1]**
   Reduction in fertility/symptoms of menopause. **[1]**
   c) Similar shape to oestrogen molecule **[1]**
   so competes for oestrogen receptor. **[1]**
   Receptor blocked so less transcription. **[1]**

Page 135

1. 1 and 3 only **[1]**
2. It reduces the time needed to cool down the DNA from 95°C. **[1]**
   a) Primers **[1]**
   b) Primers are needed on each end because DNA polymerase only works from 5' to 3'. **[1]**
   Each end has different base sequences. **[1]**
   c)

Temperature (°C)	Time at this temperature (s)	Reason for this temperature
95	30	To break the hydrogen bonds **Allowing the two DNA strands to separate**
55	45	**Cool enough to allow the primers to bind**
72	60	**Optimum temperature for Taq polymerase to work**

   One mark for each reason **[4]**
   d) One cycle = $\dfrac{135\,\text{s}}{2\frac{1}{4}\ \text{minutes}}$ **[1]**

   18 minutes = 8 cycles **[1]**
   8 cycles would produce 256 molecules **[1]**

Page 137

1. 2 and 4 only **[1]**
2. a) F, B, A, C, E, D
   One mark for each consecutive pair in the correct order **[5]**
   b) Separates DNA fragments according to their length. **[1]**
   Longer fragments will not pass as far along the gel. **[1]**
   c) 10 kb **[1]**
   d) i) $\dfrac{50\,000\,000}{500}$ **[1]**
   100 000 **[1]**
   ii) Any one from:
   Automation of the reading of fragments using fluorescence.
   New generation methods running many sequences at the same time. **[1]**

Page 139

1. a) E, B, D, C, F, A
   One mark for each consecutive pair in the correct order **[5]**
   b) The human gene contains introns. **[1]**
   Bacteria cannot remove the introns if transcribed. **[1]**
   mRNA has the introns removed. **[1]**
2. a) Restriction (endonuclease) **[1]**
   Cuts the DNA **[1]**
   Produces sticky ends **[1]**
   b) So that the sticky ends are complementary **[1]**
   allowing the gene and plasmid to join **[1]**
   c) DNA ligase **[1]**
   Joins the gene to the plasmid **[1]**
   d) Heat-treated **[1]**
   e) They possess DNA from a different species. **[1]**
   f) This is to select the bacteria that have taken up the plasmid. **[1]**
   The plasmid contains a gene for antibiotic resistance. **[1]**
   Any bacteria that are not transformed will be killed. **[1]**

Page 141

1. a) It is harmful to wildlife. **[1]**
   If burned, it may give off toxic fumes. **[1]**
   b) A bacterium **[1]**
   c) They are photosynthetic / can produce their own food **[1]**
   d) Antibiotic-resistant bacteria cannot be controlled by antibiotics. **[1]**
   They may be harmless but could transfer the resistant gene to pathogenic bacteria. **[1]**
2. a) Children 1 and 3 could have both adults as their parents. **[1]**
   Child 2 could only have one of the adults as their parent. **[1]**
   b) i) Large quantities of DNA can be produced from small samples. **[1]**
   ii) VNTRs are unique. **[1]**
3. a) Introducing a gene will not stop an existing gene from being expressed **[1]**
   so a harmful protein will still be made. **[1]**
   If a gene is producing a harmless non-functioning protein then gene therapy can introduce a working copy. **[1]**
   b) Genomic medicine **[1]**

PRACTICE PAPERS

Note: Words within answers that appear in **bold** must be used/spelled correctly.

Paper 1

01.1 α–helix **[1 mark, both needed to gain mark]**

01.2 Water **[1 mark]**

01.3 Ring drawn around either central H atom on glycine **[1 mark]**

01.4 **Any two from**:

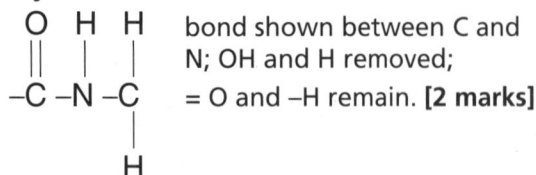

$$-C - N - C$$

bond shown between C and N; OH and H removed; = O and –H remain. **[2 marks]**

(with O double bonded to first C, H on N, H on second C, and H below second C)

01.5 **Any two from**: collagen is strong / inelastic; three polypeptide chains wound together; many hydrogen bonds between turns of the helices / between adjacent chains; **[2 marks: 1 mark for each explanation]**

02.1 Denaturation **[1 mark]**

02.2 **Any three from**: heat causes hydrogen bonds to break; polypeptide chains unfold then become tangled together; causes precipitation / polypeptides no longer soluble. **[3 marks: 1 mark for each explanation]**

02.3 Add dilute sodium hydroxide and dilute copper sulfate/biuret solution to the sample **[1 mark]**; A violet/purple colour appears if a protein is present **[1 mark]**

02.4 Endopeptidases break peptide links/bonds between amino acids **in the middle** of polypeptide chains **[1 mark]**; Exopeptidases break the peptide links of the amino acids at the **ends** of the chains **[1 mark]**; Exopeptidases form shorter polypeptides/dipeptides **[1 mark]**

03.1 Change = 84 – 71 = 13; % change = 13/71 = 18.3 **[2 marks]**

03.2 Temperature: **Thermostatically controlled** water bath / or described **[1 mark]**
pH: Buffer solution **set at optimum/appropriate pH value** (range) **[1 mark]**

03.3 Sucrose concentration ceases to be the limiting factor; At any moment all the invertase active sites are full **[2 marks]**

03.4 Domain, Phylum, Order, Saccharomyces, **[2 marks for all four correct, 1 mark for three correct, 0 mark for 0, one or two correct]**

04.1 Name of structure A is the nucleus.
Function of structure A is to store genetic material/DNA/chromatin; controls the cell's activities **[requires correct name and one function to gain 1 mark]**

04.2 Lysosomes contain digestive enzymes **[1 mark]**; Macrophage uses enzymes to digest/break down material it has ingested/engulfed **[1 mark]**

04.3 Actual diameter of X = size of image/magnification = 4/6000 = 0.00067 cm = 6.67 µm. **[2 marks for correct answer in micrometres, 1 mark if answer is 0.00067 or working shown as 4/6000]**

05.1 Stains – (acetic) orcein or toluidine blue
Any four from: cut small length of the root tips; put tips in a small volume of ethanoic acid (on a watchglass); for 10 minutes; wash the root tips then dry on filter paper; heat hydrochloric acid (to 60°C) in a water bath; transfer the root tips to the hot hydrochloric acid (and leave for 5 minutes); wash the root tips again in cold water and dry on filter paper; (use mounted needle to) remove some root tips onto a **clean** microscope slide; cut each about 2 mm from the growing root tip (discard the rest); add a small drop of stain (and leave for 2 minutes); break up the tissue with a mounted needle/seeker; cover with a coverslip and squash/description of squash.
[5 marks: 1 mark for stain, 4 marks for each description]

05.2 10/130 (allow answer between 10/130 and 14/130) = 7.7% (allow answer between 7.7 and 10.8) **[2 marks]**

06.1 Regular cell shape with cell wall, cytoplasm and central vacuole shown **[1 mark]**; Cytoplasm pulled away from cell wall in at least one place/completely detached from cell wall – cytoplasm must be labelled **[1 mark]**

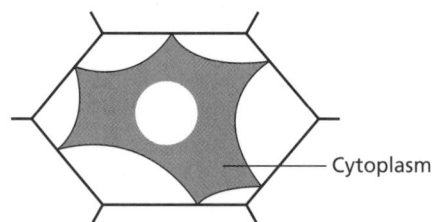

Cytoplasm

06.2 **Any two from**: external solution/soil water has a **lower** water potential; water will leave by **osmosis**; plasmolysis would lead to wilting / plant not able to absorb water for its life processes.
[2 marks: 1 mark for each explanation]

06.3 Red pigment allows cell contents/cytoplasm/protoplast to be seen more clearly **[1 mark]**

06.4 10.4% **[1 mark]**

06.5 Correct **linear** scale taking up more than half of the grid area **[1 mark]**; Correct plotting **[1 mark]**; Smooth curve/straight line through points **[1 mark, no mark for bar graph]**

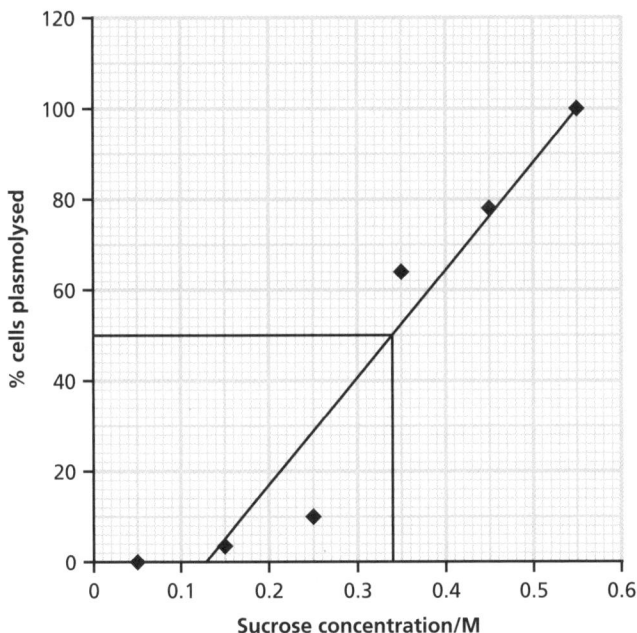

06.6 Two correct lines drawn at 50% plasmolysis **[1 mark]**; Correct molarity of onion cells – 0.32 M **[1 mark, error within plus or minus 0.02 M will be allowed]**

07.1 (Logarithmic scale) needed so that all data can be seen in the space available / difference between elephant and shrew is so great that the data would not fit on a linear scale **[1 mark]**

07.2 **At least one from**: as body mass increases, metabolic rate per gram per hour decreases; for body masses greater than 1.05 kg, the differences between metabolic rates are less pronounced.
At least one from: smaller animals have a greater surface area for their mass than larger ones; therefore, they lose heat at a faster rate; and require more energy to maintain a constant core temperature; smaller animals, e.g. shrews, are more active as they need to eat a proportionally higher mass of food each day. **[3 marks: 1 mark for each explanation]**

07.3 **Any two from**: protists have a large surface area to volume ratio, or reverse argument (ORA); shorter diffusion pathway to all parts of organism (ORA); diffusion an adequate method for exchanging gases in protists / larger organisms require specialised organs to transport respiratory gases long distances.
[2 marks: 1 mark for each explanation]

08.1 0.3 seconds **[1 mark]**

08.2 Arrow at 0.42 seconds where the lines cross **[1 mark]**

08.3 **Any two from**: atrioventricular valves are closed; to prevent backflow of blood; into atria. **[2 marks: 1 mark for each explanation]**

08.4 0.11 dm^3 per minute **[2 marks, but if incorrect answer 7.15 divided by 65 gains 1 mark]**

08.5 **Any three from**: blood is transported more quickly; more oxygen taken up at the lungs / more carbon dioxide excreted at the lungs; more oxygen reaches the muscles; more glucose reaches the muscles; so muscles contract more effectively. **[3 marks: 1 mark for each explanation]**

09.1 **Any two from**: haemoglobin has a high affinity (attraction) for oxygen (at this partial pressure); it becomes highly saturated with oxygen; sigmoid or 'S'-shaped dissociation curve/properties of haemoglobin mean that a small change in partial pressure causes a massive loading of oxygen. **[2 marks: 1 mark for each explanation]**

09.2 **Any two from**: the dissociation curve is moved to the right/Bohr shift; more oxygen is 'off-loaded' to the muscle; carbon dioxide lowers the affinity of the haemoglobin for oxygen. **[2 marks: 1 mark for each explanation]**

09.3 The curve should be drawn **to the left of the** normal line with the **same** origin **[1 mark]**

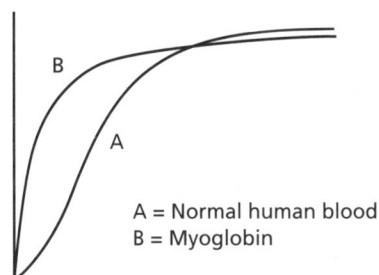

A = Normal human blood
B = Myoglobin

09.4 Myoglobin remains bound to oxygen when whale is submerged **[1 mark]**; Oxygen store in muscle can be used to supply the needs of respiration while under water **[1 mark]**

10.1 Plot A species diversity = 3.94 **[1 mark]**; Plot B species diversity = 2.70 **[1 mark]**
Working as follows:
Plot A: $N = 46$, total $n(n – 1) = 526$, therefore $(46 \times 45)/526 = 3.94$
Plot B: $N = 42$, total $n(n – 1) = 640$, therefore $(42 \times 41)/640 = 2.69$

10.2 **Plot A has a greater diversity** as it has been left uncultivated longer than Plot B **[1 mark]**; Greater opportunity for new species to colonise in Plot A / species have more time to multiply in number or reproduce / habitat has more time to recover from effects of cultivation or eradication of wild species **[1 mark]**

10.3 For Rosebay willow herb, the difference in number is probably due to chance/probability – the likelihood of this is more than 5%/0.05 **[1 mark]**; For all the other species, the difference in number is unlikely to be due to chance **[1 mark]**; Because P for these species < 0.05, which is significant/is lower than 5%/0.05 **[1 mark]**

11 **Any two from:** ATP is formed from a molecule of ribose and adenine; and three phosphate groups; **[these 2 marks can be gained from a suitable molecular diagram]**; ATP is synthesised by the **condensation** of ADP and Pi; catalysed by the enzyme **ATP synthase.**

Any three from: ATP can release its energy as a small 'packet' by **hydrolysis**; controlled by the enzyme **ATP hydrolase**; inorganic phosphate released; can be coupled to energy-requiring reactions within cells / suitable example, e.g. synthesis of large molecules, active transport; ATP can be used to phosphorylate other compounds, often making them more reactive / reference to photosynthesis or respiration. **[5 marks: 1 mark for each explanation]**

12 **At least one from:** transpiration occurs in **xylem** (dead cells); transpiration flow carries **water** and **mineral ions**; movement of water in transpiration due to combination of **root pressure** and **cohesion-tension**; reference to **capillarity** contributing to **cohesion** (water molecules have an attraction for each other so when one water molecule moves others move with it); water molecules are attracted to the sides of the vessels, which pulls the water upwards (adhesion); transpiration causes a very negative water potential in the mesophyll of the leaves / water in the xylem is of higher water potential.

At least one from: translocation occurs in **phloem**; translocation occurs via mass flow/ cytoplasmic streaming; translocation involves transport of carbohydrate/sucrose and amino acids; reference to companion cells/possession of cytoplasm and mitochondria/cells are living. **[6 marks: 1 mark for each explanation]**

13 **At least one from:** vaccines consist of attenuated pathogens/non-harmful antigens; antigens trigger B-lymphocytes/humoral immune system to produce antibodies; they stimulate the production of memory cells; so that the person develops **active artificial immunity.**

At least one from: herd immunity – explanation of herd immunity, e.g. the vaccination of a significant proportion of a population (or herd) makes it difficult for a disease to spread; because there are so few susceptible people left to infect. **[4 marks: 1 mark for each explanation]**

Paper 2

01.1 To release carbon dioxide for fixation in light-independent reaction/photosynthesis **[1 mark]**

01.2 To ensure light-dependent reactions had stopped / to analyse compounds formed at a particular time **[1 mark, 'to stop photosynthesis/kill algae' insufficient]**

01.3 To release compounds from cells / to break up cells / to enable compounds to be identified **[1 mark]**

01.4 A: glycerate 3-phosphate (first compound to be formed in light-independent reaction and therefore first to contain radioactive carbon) **[1 mark]**
B: triose phosphate (formed later, hence not seen on 5-second chromatogram) **[1 mark]**

01.5 54.5% **[2 marks]** but $(6.8 - 4.4)/4.4 \times 100$ **[1 mark]**

01.6 glycerate 3 phosphate continues to be formed from ribulose bisphosphate; **[1 mark]**; but cannot be converted to triose phosphate; **[1 mark]**

01.7 No triose phosphate formed as no reduced NADP or ATP available **[1 mark]**; therefore no compound starting point available for formation of more ribulose bisphosphate **[1 mark]**

01.8 Rubisco / RuBP carboxylase / ribulose bisphosphate carboxylase **[1 mark]**

02.1 Nitrification **[1 mark]**

02.2 Atmospheric nitrogen converted to ammonium **[1 mark, 'ammonia' will not be accepted]**; Nitrogen in ammonium needed for amino acids/ protein/nucleic acids **[1 mark]**

02.3 **Any two from:** water fills air spaces / causes lower oxygen content; anaerobic conditions encourage growth of **denitrifying bacteria**; nitrate converted back to molecular/atmospheric nitrogen. **[2 marks: 1 mark for each explanation]**

02.4 Insects trapped/digested are a source of protein **[1 mark]**; Protein is broken down in body of plant to release nitrogen/nitrate **[1 mark]**

02.5 To incorporate into/build ATP molecule/build DNA to prevent poor growth/poor quality fruit/ blue or green leaves/phospholipids for cell membranes **[1 mark]**

02.6 $r = 0.84$ **[3 marks for correct answer, but correct numerator – 243 910 – will gain 1 mark, and correct denominator – 289 204, i.e. 243 910/289 204 – will gain 1 mark]**

02.7 'Yes' – 0.84/answer to 'Explanation' is a significant correlation **or** if r is miscalculated at <0.5. 'No', r is not a significant correlation, will be allowed.
However, data may not be valid due to small sample size/timing of sample/phosphate may have entered waterway at different points / algae may not have had time to respond / other mineral nutrients not accounted for, etc. **[1 mark: No mark for 'Yes' or 'No', must be justified.]**

02.8 **Any three from**: growth of algal blooms on surface; blocks off sunlight; less photosynthesis; plants/algae die; aerobic bacteria break down dead plant material; oxygen absorbed from water; animals, e.g. fish, die **due to lack of oxygen** ['algae kill animals' will not be allowed]. **[3 marks: 1 mark for each description]**

03.1 There is no significant difference between the number of woodlice found in dark and light / woodlice have no preference for light or dark conditions **[1 mark]**

03.2 **Any two from**: woodlice of the same age; constant room temperature; constant humidity; uniform/lack of vibration/noise; controlled/no handling of woodlice/use of brush or similar to avoid trauma; removal of olfactory/chemical secretions from woodlice (by using fresh choice chamber/base) when performing replicates. **[2 marks: 1 mark for each variable]**

03.3

	Light	Dark
Observed results (O)	39	61
Expected results (E)	50	50
$(O–E)^2$	121	121
$(O–E)^2/E$	2.42	2.42

Correct value of χ^2 is 4.84 **[1 mark for correct value of χ^2; 1 mark for correct totals in raw data table; 1 mark for correct completion of chi-squared table]**

03.4 If χ^2 correct, then 4.84 is greater than critical value **[1 mark]**; Therefore null hypothesis is rejected / probability that difference in distribution is due to chance is <0.05/5% **[1 mark]**

03.5 Directional response **[1 mark]**; Away from (touch) stimulus **[1 mark]**

04.1

gametes

RrYy × RrYy

(RY) (Ry) (rY) (ry)   (RY) (Ry) (rY) (ry)

[1 mark, all must be correct, can be in any order]

	RY	Ry	rY	ry
RY	RRYY	RRYy	RrYY	RrYy
Ry	RRYy	RRyy	RrYy	Rryy
rY	RrYY	RrYy	rrYY	rrYy
ry	RrYy	Rryy	rrYy	rryy

[2 marks for completely correct table, every incorrect genotype subtracts 1 mark]

04.2 9 round and yellow:3 green and round: 3 yellow and wrinkly:1 green and wrinkly **[1 mark, all must be correct]**

04.3 **Any two from**: linkage between chromosomes; numbers in sample not high enough to show expected ratios; loci of genes may be close together on the same chromosome. **[2 marks: 1 mark for each reason]**

04.4 Denise's possible genotypes: X^HX^h or X^hX^h **[1 mark]**

Explanation: Johnny is X^HY, so he is responsible for Bill's Y chromosome (Y chromosomes do not carry a blood clotting gene) **[1 mark]**; Working backwards, Bill has haemophilia so Denise must have at least one X^h **[1 mark]**

05.1 **Any three from**: the pancreas fails to produce enough insulin; after a meal when blood glucose concentrations increase, the level remains high; high blood glucose causes hyperglycaemia; kidneys cannot reabsorb glucose, causing glucose presence in urine. **[3 marks: 1 mark for each description]**

One symptom from: dehydration; loss of weight; increased risk of infection; genital itching; lethargy. **[1 mark]**

05.2 Advantages, **any two from**: better/more precisely controlled blood glucose levels / less likely to become hyper- or hypo-glycaemic; the system simulates the body more precisely; no need to manually check glucose levels by blood testing; improved quality of life.

Disadvantages, **any two from**: more expensive / may not be available for everybody on the NHS; possible malfunctions in control unit could lead to incorrect insulin dosages; closed-loop system – cannot use manual injections as a back-up if problems occur. **[4 marks: 1 mark for each advantage, 1 mark for each disadvantage]**

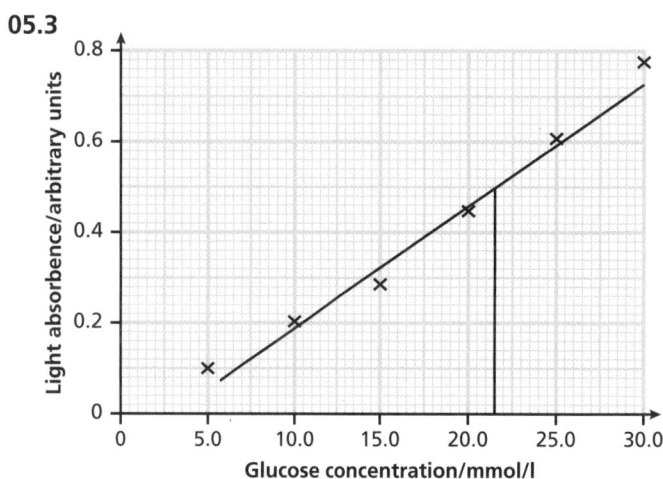

05.3

Straight line (of best fit) to be drawn on graph, 21.5 mmol/l **[1 mark, error of plus or minus 0.5 will be accepted if answer incorrect]**

05.4 Patient has blood glucose levels significantly higher than that of a normal/average/healthy adult **[1 mark]**; Therefore, the patient is likely to have diabetes **[1 mark]**

06.1 $q^2 = 0.3$ **[1 mark]**
$q = 0.55$ **[1 mark]**

06.2 $q = \sqrt{0.3} = 0.55$, then from $p + q = 1$, $p = 1 - 0.55 = 0.45$ **[1 mark]**; then Heterozygote frequency = $2pq = 2 \times 0.45 \times 0.55 = 0.5$ **[1 mark]**; In a population of 3000, this would be 1500 heterozygote mice **[1 mark]**

06.3 **Any two from**: there must be no immigration or emigration; no mutations; no selection (natural or artificial); must be true random mating; all genotypes must be equally fertile. **[2 marks: 1 mark for each assumption]**

07.1 **Any four from**:
Allopatric speciation takes place after geographical isolation; Named example, e.g. the rising of sea level splits a population of animals formerly connected by land to create two islands; Mutations take place so that two groups result in different species.
Sympatric speciation takes place through genetic variation; In the same geographical area; Mutation may result in reproductive incompatibility; Named example, e.g. a structure in birds may lead to a different song being produced by the new variant; This may lead to the new variant being rejected from the mainstream group; Breeding may be possible within its own group of variants **[4 marks: 1 mark for each explanation, one of which must be an example]**

07.2 Data concerning genetic sequencing would contain limited sequences of bases/codons/triplets that code for specific genes **[1 mark]**; The genome is the entire genetic code of an organism, therefore will contain the sequences that code for thousands of genes **[1 mark]**

07.3 **Any two from**: more closely related species will share a greater proportion of CNVs; common ancestor(s) will share CNVs with more than one line of descendants; evidence from these studies can be used to reinforce observations from other phenotypic variations and, thus, establish evolutionary trees.
[2 marks: 1 mark for each suggestion]

07.4 **Any two from**: presence of certain methyl groups determine which genes are switched on; increased methylation reduces transcription of genes; therefore common phenotypic traits can be used to assess how closely certain species are related; presence of certain proteins (and their switched-on genes) gives information about common ancestry.
[2 marks: 1 mark for each property]

07.5 **Any five from**:
Natural selection: beak size/shape/type coded for by particular alleles/genes; differences in beak size caused by mutations; range of alleles/genes produces range of beak types within a population; conditions/food availability in different habitats/islands acts as a **selective pressure**; the better adapted variants within the common ancestral population survive, e.g. greater amount of larger nuts favours a larger beak; these individuals are more likely to reproduce and pass on these **alleles** (NOT genes) to the next generation; allele for advantageous trait becomes more common in a population.
Speciation: emergence of new species takes place when isolation (geographic or reproductive) occurs, e.g. separation by ocean flooding (production of islands); over time separated populations can no longer interbreed. **[5 marks: 1 mark for each explanation]**

08.1 Bacteria can no longer interfere with growth (caused by cytokinins and auxins), therefore no tumours are formed **[1 mark]**

08.2 Restriction endonucleases **[1 mark]**

08.3 Antibiotic kills those bacterial cells that have not taken up the resistance DNA coding **[1 mark]**; Hence, plant cells can only be infected with bacteria that have taken up the desired gene for vitamin A **[1 mark]**

08.4 To allow bacterial plasmids to be taken up by the plant cell **[1 mark]**

08.5 No retinol available for pigment in **rods** **[1 mark]**; Rod cells are responsible for night vision **[1 mark]**

08.6 Light falls on photosynthetic pigments in rod **[1 mark]**; Chemical change in receptor – rhodopsin $\rightarrow$ opsin + retinal **[1 mark]**; Opsin opens ion channels in the cell surface membrane **[1 mark]**; Causing influx of sodium ions **[1 mark]**; Depolarisation of axon initiates action/generator potential **[1 mark]**

Paper 3

01.1 C: glycerate 3-phosphate/G3P; D: acetyl co-enzyme A **[1 mark for both correct]**

01.2 NADH needs to be oxidised / hydrogen taken away so that it can continue to accept hydrogen in glycolysis **[1 mark]**

01.3 Plants: (hydrogen accepted by) ethanal pyruvate (and) converted to ethanol **[1 mark]**
Animals: (hydrogen accepted by) pyruvate (and) converted to lactate **[1 mark]**

01.4 Build up of lactate [1 mark]; Lactate is a poison/toxic/damages muscle tissue, causes muscle cramp [1 mark]

02.1 Arg **Ile Ser Gly Arg Leu Cys Asp Thr Pro** [all correct for 1 mark]

02.2 2 **Hydrogen bonds** between the two chains **break** and the two strands separate [1 mark].
3 Each **complementary strand** then acts as a **template** to build its opposite strand **from free DNA nucleotides** [1 mark]
4 The enzyme **DNA polymerase joins the nucleotides together** (this process results in the production of two identical copies of double-stranded DNA) [1 mark]

02.3 **Any three from:** breast tissue cells without receptors no longer require oestrogen to trigger mitosis/cell division; **mitosis** occurs spontaneously/is not controlled/not stopped; signals from neighbouring cells ineffective/cancer cells do not respond to neighbouring cells' signals; cancer cells divide to produce large mass/tumour. [3 marks: 1 mark for each explanation]

02.4 **Any four from:** drug/inhibitor prevents transcription of signal molecule gene; signal molecule/polypeptide no longer produced in ribosome; tumour produces fewer/no blood cells; tumour receives less/no nutrients/oxygen; required for growth of tumour/production of new cancer cells. [4 marks: 1 mark for each explanation]

03.1 **Any four from:** SSRI molecule blocks serotonin transport molecule; resulting in more serotonin being present in synaptic cleft; therefore, more serotonin binds with receptor sites on post-synaptic membrane; which triggers more nerve impulses/action potentials in the neurone/has an excitatory response; this stimulates centres of the brain associated with mood. [4 marks: 1 mark for each explanation]

03.2 Pre-cursor increase results in greater amount of dopamine [1 mark]; Increased dopamine excites areas of the brain involved in motor control/muscle movement [1 mark]

03.3 **Any two from:** vesicles release more dopamine; less dopamine reabsorbed into pre-synaptic membrane; more dopamine available to trigger action potentials impulses in post-synaptic membrane. [2 marks: 1 mark for each explanation]

03.4 **Any two from:** a single action potential may arrive at a synaptic knob; there may not be enough transmitter molecules being secreted into a cleft to cause an action potential to be generated; a series of action potentials arrive at the synapse to build up transmitter substances to reach the threshold; the neurone will now send an action potential. [2 marks: 1 mark for each explanation]

04.1 Antigen binding site [1 mark]

04.2 **Any two from:** binding site **specific** to **antigen** on pathogen (outer membrane); binds/locks on to the pathogen; pathogens clumped together/agglutination (complement); pathogen's toxin neutralised. [2 marks: 1 mark for each description]

04.3 Measure colour density / use spectrophotometer or colorimeter [1 mark]; The amount of colour change is proportional to the amount of antigen (venom) present in the test sample [1 mark]

04.4 Test different snake venom molecules/proteins with **the same** antibody/IgG [1 mark]; A positive reaction/complex formed indicates the antibody/anti-venom is effective [1 mark]

04.5 **Any two from:** venom disrupts/breaks bonds holding polypeptide chain/fibrinogen in blood together; 3D/tertiary structure lost; molecule unravels/globular structure becomes long chain; chains tangle/link together and precipitate/coagulate. [2 marks: 1 mark for each suggestion]

04.6 **Any two from:** a food **vacuole** is formed around the pathogen; hydrolytic **enzymes** from **lysosomes**/lysosomes join with food vacuole/reference to phagosome; enzymes digest/destroy pathogen. [2 marks: 1 mark for each description]

05.1 (Low) density [1 mark]

05.2 **Any two from:** plastic fibre content is much greater in sub-tidal regions than in estuarine or sandy regions; estuarine fibre concentration is greater than sandy; error bars do not overlap, so this difference is significant.
[2 marks: 1 mark for each conclusion]

05.3 **Maximum of one from:** production of synthetic fibres shows a general trend of increase from the 1960s to the 1990s; there have been temporary, small dips in production on three occasions during this time period.
Maximum of two from: concentration of fibres greatest in 1980s; however – error/standard deviation bars in the two decades overlap considerably, so this might not be significant/any quoted comparison of data, e.g. fibre concentration in 1980s 0.045 per m^3 compared with 0.035 per m^3 in 1990s.
[3 marks: 1 mark for each explanation]

05.4 To see if different organisms had different uptake rates [1 mark]; There is a range of feeding types [1 mark]

05.5 **Any two from:** plastic fibres are not toxic in and of themselves; no evidence from this study that shows toxic substances are given off the fibres; there is no evidence for further accumulation of fibres further up the food chain/in consumers of these organisms. [2 marks: 1 mark for each point made]

05.6 Any two from:

bioaccumulation/biomagnification;
concentration of toxins increases further up the
food chain where vertebrates tend to be found;
toxins are persistent.
[2 marks: 1 mark for each explanation]

05.7 Stated concentrations of PCBs are higher in
liver/standard deviations for liver are larger
[1 mark]; But standard deviations overlap in each
organ category **[1 mark]**; Therefore may not be
significant **[1 mark]**

06 Essay 1, ideal response, which would achieve 25 marks:

Global warming causes a rise in average ambient temperatures. This masks the effects in specific climates and particular species, as the temperature changes in some situations can be profound and varied. Mean temperatures in some latitudes may actually go down due to the effect on weather systems. The effects on organisms can be grouped into those that have a physiological basis and those that operate on an ecological or macro level, although the two are very much connected.

In terms of physiology, a change in ambient temperature has a marked effect on ectotherms (organisms whose core temperature depends on the environment), but even endotherms (e.g. mammals and birds) will be affected to a certain degree, despite them having some control of body temperatures via homeostasis. In general terms, an organism's rate of metabolism will increase due to the effect of increased temperature.

Higher temperatures mean that the kinetic energy of molecules is greater. Therefore, enzyme activity increases (up to an optimum point) as there are more collisions per second between enzymes and their substrate molecules. This means that all reactions in the bodies of organisms will be affected, from digestion to respiration.

In plants, rate of photosynthesis is known to increase with temperature as the reactions that constitute the process rely on enzyme control. This increase is limited as, at temperatures above 40 °C, other factors such as carbon dioxide concentration or light intensity act as limiting factors.

As respiration rate also increases with temperature, then activity of ectotherms is affected. Such organisms may need to migrate or alter their behaviour to withstand extremes of hot or cold.

Decomposition has the process of respiration at its centre, thus bacteria and fungi will have increased metabolism, meaning rates of decay increase. This has implications for ecological niches (see later).

Movement of materials across cell membranes, whether they are molecules, ions or atoms, will also increase in rate as the processes depend on kinetic energy. Higher kinetic energy increases rates of diffusion in particular, which has implications for gaseous exchange and transpiration. In higher animals, increased temperature causes a shift to the right of the oxygen dissociation curve of haemoglobin. This means that more oxygen is unloaded due to a decreased affinity for the haemoglobin.

In water, availability of oxygen will be lower as high temperatures increase the release of gases in solution. This may affect the survival and distribution of aquatic animals sensitive to oxygen concentration, e.g. stonefly nymphs and fish.

In general, ectotherms, e.g. reptiles, will be more active during the daytime. Behavioural adaptations such as exposure to the heat of the Sun will keep their metabolism at optimum conditions, which, in turn, makes them more competitive with endotherms. Other physiological effects in the bodies of endotherms include the effect on nervous transmission in receptors and neurones. In terms of thermo-regulation, the hypothalamus is more likely to trigger impulses to effectors, which increase sweating and vasodilation.

In plants, transpiration is accelerated at higher temperatures because evaporation, and hence diffusion of water from the spongy mesophyll cells in the leaf (and its loss through the stomata), increases. This in turn draws more water up through xylem cells as a result of the cohesion of water molecules. In high-temperature regimes, xerophytic adaptations become more predominant, such as leaves having reduced surface area to volume ratio (like cacti), or rolled-up leaves and sunken stomata. Higher temperatures are therefore acting as a selective pressure on plants.

All the above physiological effects influence the wider behaviour and ecology of organisms. For example, microbe decomposers (i.e. saprophytes) increase in number as the rates at which cells divide is increased. In terms of the carbon cycle, more carbon dioxide is released as aerobic respiration goes up. This in turn increases global warming as it is a greenhouse gas.

Unpredictable temperature changes cause migration or extinction of species. This will disrupt food webs due to new mixes of species and different interactions in terms of competition and predation. Populations that were once cohesive will become more isolated geographically, which in turn drives speciation and therefore evolution (see later).

In plants, temperature changes reduce biodiversity as fewer species are adapted for hot, dry climates. Despite increases in photosynthesis rate, the loss of so many plant species will reduce productivity, meaning fewer habitats and food for the animals that depend on them. As population numbers decrease this also has the effect of shrinking the gene pool. The lower range of genetic types make populations more vulnerable to environmental conditions and increase the frequency of harmful recessive alleles.

Selection pressures on higher animals would cause less hair and a leaner body with a higher surface area to volume ratios for rapid heat loss. So, ultimately, increased global temperatures will markedly affect the direction of evolution in the whole biosphere.

Essay 2, ideal response, which would achieve 25 marks:

Water is a major component of all cells. Movement of substances dissolved or suspended in this liquid is a vital necessity for all organisms. Single-celled organisms such as protists have no mass flow system and therefore rely on one or more of the following: diffusion, osmosis, active transport and facilitated diffusion. It should also be recognised that these processes may work alongside each other in the multicellular state.

All organisms are bounded by a phospholipid bilayer through which materials must pass. Moreover, eukaryotic cells contain membrane-bound organelles such as mitochondria and ribosomes. These present another barrier to movement of substances but also allow compartmentalisation of materials inside organelles, and the ability to carry out chemical reactions isolated from other interfering agents. Cell membranes are selectively permeable, meaning that they allow some substances through but not others.

Diffusion and its other form – facilitated diffusion – account for much of the movement across membranes and can occur directly through the membrane's lipid bilayer provided the substances are either lipid-soluble (i.e. hydrophobic), e.g. steroids, or very small, e.g. water, oxygen or carbon dioxide.

Diffusion is the movement of particles in a liquid or gas from a region of high concentration to one of low concentration. An example in the human body is gas exchange in the alveoli. Adult human lungs have a total area of around 100 m². This illustrates an important principle that diffusion is far more effective when there is a high surface area to volume ratio. The folded nature of the alveoli provides this. In addition, the single layer of cells in the alveolar membrane and capillary endothelium provide a short diffusion pathway, which also increases diffusion rate.

Facilitated diffusion is the movement of substances across a membrane through trans-membrane protein molecules spanning the whole phospholipid bilayer. The transport proteins are largely specific for one molecule, which means that substances can only cross a membrane containing the correct protein.

As with simple diffusion, the process is passive; therefore, no energy is expended and substances move down their concentration gradients. There are two kinds of transport protein: channel proteins, which form a water-filled pore or channel in the membrane; and carrier proteins that possess a binding site for a specific solute and repeatedly change between two states. This makes the site alternately open on opposite sides of the membrane. The substance will bind on the side where it is at a high concentration and be released where it is at a low concentration. Examples of substances that use this mode of transport include glucose and amino acids.

Glucose is a hydrophilic monosaccharide and cannot diffuse directly through the hydrophobic membrane. Specific extrinsic glucose channel proteins have a complementary shape that allows glucose to enter the cytoplasm by facilitated diffusion down a concentration gradient. This absorption of glucose is stimulated by the presence of insulin produced from the pancreatic cells.

Occasionally, carrier proteins have two binding sites and can carry two molecules at once. This is called cotransport, e.g. the sodium/glucose cotransporter found in the small intestine.

A special case of diffusion is osmosis. The contents of cells contain solutions of many different solutes, and each solute molecule attracts water molecules in a hydration shell. The more concentrated the solution, the more solute molecules there are in a certain volume, and consequently the more water molecules included in the hydration shell. This means that there are fewer free water molecules to diffuse easily across the membrane of a cell. The overall effect of this is that the net movement of water occurs down its concentration gradient, or in other words, from a dilute solute concentration to a more concentrated one. Putting it another way, water moves down a water potential gradient.

There are many examples where this occurs between plant cells, e.g. in the mesophyll of the leaf. Here, the dilution of cell contents as water moves into the cell raises the water potential, thus enabling water to move on into the next cell and so on. In the mammalian kidney, osmosis accounts for the re-absorption of water from the collecting duct. Here, sodium ions in the medulla act as the solute, effectively drawing the water out of the collecting duct and back into the capillaries.

Active transport is where pumping of substances takes place across a membrane via a trans-membrane protein pump molecule. The protein binds a molecule of a substance to one side of the membrane, then changes shape and releases it on the other side. These proteins are very specific; therefore, a different protein pump exists for each transported molecule. The protein pumps are ATPase enzymes. The splitting of ATP into ADP and phosphate as catalysed by this enzyme releases energy to change the shape of the protein pump. This means that active transport requires the release of energy and is the only transport mechanism that can transport substances against a concentration gradient. An example of active transport is in root hair cells where mineral ions such as nitrate and potassium ions are absorbed from the soil against a concentration gradient.

Movement of molecules within a cell occurs from organelle to organelle, e.g. in respiration, pyruvate produced from glycolysis, moves from the cytoplasm into the mitochondrial matrix. Here, it is a reactant in the Krebs cycle and produces reduced NAD and FAD co-enzymes. These then diffuse to the cristae where they pass their electrons down an electron transport chain in a series of redox reactions. Similar transport of hydrogen ions within chloroplasts occurs during the light independent reaction of photosynthesis. This process is termed chemiosmosis.

Larger molecules, such as messenger RNA, can migrate from the nucleus to the ribosomes in order to manufacture polypeptides in the process of translation. Although larger than the water molecule, mRNA is still small enough to diffuse out through the nuclear pores and bind to ribosomes on the rough endoplasmic reticulum.

Further specific examples of movement across membranes are nervous conduction and the setting up of action potentials in receptors and neurones. Here, the small potential is generated by the influx of sodium ions through a sodium channel. Additionally, synaptic transmission requires the diffusion of a neurotransmitter between neurones at the nerve junction.

We can therefore see that movement of substances is a basic requirement of life, and can be achieved by varied means that result in a high degree of control.

Essay 1: guide to the allocation of marks

Specification reference	Topic area
3.1.4	Enzymes and their function [2 marks]
3.2.3	Movement across cell membranes [2 marks]
3.3.1	Effect of SA/Vol ratio [1 mark]
3.3.2	Gas exchange and oxygen concentration in water [1 mark]
3.3.4.1	Temperature's effect on haemoglobin [1 mark]
3.3.4.2	Transpiration and diffusion rates of water vapour from leaves [2 marks]
3.4.4	Genetic diversity and adaptation [2 marks]
3.4.6	Biodiversity [2 marks]
3.5.1	Rates of photosynthesis [2 marks]
3.5.2	Rates of respiration [2 marks]
3.5.4	Effect of temperature on rate of decomposition [1 mark]
3.6.2.1	Effect of temperature on speed of conductance along neurones [1 mark]
3.6.4.1	Homeostasis and control of body temperature [2 marks]
3.7.3	Speciation (geographic isolation) [1 mark]
3.7.4	Migration of organisms and issues relating to the conservation of species and habitats [3 marks]

Essay 2: guide to the allocation of marks

Specification reference	Topic area
3.1.7	Water [1 mark]
3.1.8	Inorganic ions [2 marks]
3.2.3	Transport across cell membranes [6 marks]
3.3.1	Surface area to volume ratio [1 mark]
3.3.2	Gas exchange [2 marks]
3.3.3	Digestion and absorption [2 marks]
3.5.1	Photosynthesis [2 marks]
3.5.2	Respiration [2 marks]
3.6.1.2	Receptors [1 mark]
3.6.2.1	Nerve impulses [2 marks]
3.6.2.2	Synaptic transmission [1 mark]
3.6.3	Skeletal muscles [1 mark]
3.6.4.2	Control of blood glucose concentration [1 mark]
3.6.4.3	Control of blood water potential [1 mark]

Notes

Notes